ARM Microcontroller Interfacing

Hardware and Software

Warwick A. Smith

Elektor International Media BV
p.o box 11
6114 ZG Susteren
The Netherlands

British Library Cataloguing in Publication Data
A catalogue record for this book is available from the British Library

ISBN 978-0-905705-91-0

Prepress production: Hans van de Weijer
First published in the United Kingdom 2010
Printed in the Netherlands by Wilco, Amersfoort

109021-UK

Table of Contents

Introduction

This book has been written to empower the reader with knowledge both to interface various hardware devices to ARM microcontrollers and write the software in the C programming language to control these hardware devices.

Although this book targets ARM microcontrollers and a certain ARM7 microcontroller from Atmel, the interfacing circuits and software programming principles apply to many other microcontrollers as well.

Only free or open-source programming tools and software are used throughout the book.

Target Audience

This book has been written for electronic engineers, students and hobbyists and is intended for use:

1) As a hardware interfacing and software reference book for microcontrollers in general, but specifically targets ARM microcontrollers.

2) For anyone who has learned the C programming language and would like to learn more about interfacing various hardware devices to ARM microcontrollers and how to program these devices.

3) As a follow on book to "C Programming for Embedded Microcontrollers" from Elektor that teaches the C programming language on embedded ARM systems. This book now provides more hardware and software examples.

Prerequisites

The reader must be familiar with programs in the C programming language as the software examples are all written in C.

Basic electronics and some knowledge of microcontrollers is required.

Introduction to the ARM Microcontrollers used in this Book

This book uses the **AT91SAM7S256 ARM7** microcontroller from Atmel as its primary microcontroller in interfacing examples. This microcontroller is one from the **AT91SAM7S family** of microcontrollers. All of the original members of this family are pin compatible i.e. the AT91SAM7S512, ‘256, ‘128 and ’64, varying only in the amount of Flash and SRAM memory on chip. Some newer members were added to this family of microcontrollers that have less memory and/or diminished features, fewer pins and are therefore not pin compatible with the original family.

These microcontrollers are “stand alone” microcontrollers as they do not have an external bus for interfacing to external memory in which programs can be run (although serial memory for data storage can be added to the TWI or SPI serial busses).

AT91SAM7S256 basic features:

1) Contains a 32-bit ARM7 microprocessor core and can therefore be programmed using open-source C language programming tools – the GNU ARM toolchain. Commercial tools are also available from several vendors.

2) Have 256k Bytes of on-chip Flash memory and 64k Bytes of on-chip SRAM memory. Programs can be run from both of these memories and they can both store data.

3) Programs can be loaded onto these microcontrollers via a direct USB connection to a PC, a direct RS-232 connection to a PC or via a JTAG ICE emulator using free software from Atmel running on the PC. No special programmer is required and the microcontroller can be programmed and re-programmed while it is soldered to the circuit board (ISP – In System Programmable).

4) Have integrated peripheral devices such as ADC, timers, general purpose I/O controller, SPI port, UARTs, etc. that are multiplexed with other devices that share the same I/O pins. E.g. a pin on the microcontroller can be set up as an input, or an output, or be connected to one of two internal peripheral devices. What device is connected to what pin is configured in the software program. The pins are individually configurable.

5) Operate at 3.3V I/O voltage. The core of the microcontroller operates at 1.8V, but a regulator on the chip derives this voltage from the 3.3V supply.

6) Have a JTAG port for single step debugging and loading programs using an external hardware device – the JTAG ICE.

7) Start running without an external oscillator. When power is applied to the microcontroller, it uses an internal RC oscillator to start running the program. The program must enable the main oscillator that runs from an external crystal.

8) Can run up to a speed of 55MHz, but are usually operated at 48MHz using an external 18.432MHz crystal. The internal clock generation circuitry derives the 48 MHz from the 18.432MHz input clock. Running the microcontroller at 48MHz generates the correct clock speed for operating the USB port.

Hardware Requirements

An evaluation board containing a member of the AT91SAM7S family of microcontrollers and preferably the AT91SAM7S256 microcontroller from Atmel is recommended for use with this book, for example the AT91SAM7S-EK evaluation kit from Atmel as described in chapter 1.

Many of the circuits in this book can be built by connecting wires from an evaluation board to a breadboard and assembling the circuit on the breadboard.

In addition to an evaluation board, a JTAG ICE is recommended for debugging. A serial port on the PC or a USB to serial port converter is also required.

Summary of hardware requirements:

1) AT91SAM7S256 evaluation board.

2) A USB to RS-232 port adapter if a RS-232 port is not available on the PC.

3) A JTAG ICE (In-Circuit Emulator) is highly recommended but not essential – e.g. the Atmel SAM-ICE from **www.atmel.com** or Amontec JTAG key from **www.amontec.com**.

4) An electronic breadboard and/or vero-board on which to build the circuits.

Additional electronic components will be required as described in each chapter.

Software Requirements

The YAGARTO GNU ARM toolchain from **www.yagarto.de** is used for writing and compiling the C programs in this book. This toolchain runs on Windows XP and Vista. It can be downloaded for free. Alternately the Sourcery G++ Lite GNU ARM toolchain from **www.codesourcery.com** can be used. This toolchain from CodeSourcery is also available for free.

The software example programs and template files used in this book can be downloaded from the Elektor website at **www.elektor.com**.

Chapter 1 contains more information on software and hardware setup and debugging.

Advice on Learning a New Microcontroller

The most important source of information when learning a new microcontroller is the microcontroller datasheet. Information can be obtained from the datasheet that will tell you how much memory the microcontroller has, how it boots up, what voltage it operates at, what integrated peripherals it contains and more.

Typical steps to follow when learning a new microcontroller are:

1) **Read the datasheet** – Datasheets are long and contain a lot of information that you might not need to read right away. For example you probably won't need every integrated peripheral device that is on the microcontroller, so skip those chapters.

2) **Evaluate the software tools** – Determine what software toolchains are available. A microcontroller is of no use unless you can program it. For the hobbyist and student, free software tools are usually a necessity so the availability of GNU tools for the microcontroller are essential.

3) **Loading programs to the microcontroller** – Determine how a program is loaded into the microcontroller. In-system programmable (ISP) microcontrollers are a good choice to cut costs, especially if the microcontroller can be programmed through a serial or USB port without any additional hardware required.

4) **Hardware tools** – Have a look at what hardware tools such as programmers or in-circuit emulators are available and if they fit your budget.

5) **Evaluation board** – Buy an evaluation board so that you can start writing programs for the new microcontroller. Not only must the microcontroller be suitable for your project, but the software toolchain and emulator must be of suitable quality. The microcontroller and programming tools can only be properly evaluated by writing programs and running them.

6) **Build Project** – If the microcontroller and tools are suitable, either use the evaluation board, a pre-built board containing the chosen microcontroller or design and build your own microcontroller board for use with your project.

It may be tempting to build your own microcontroller board without buying an evaluation board. This is possible, but the safest way is to buy an evaluation board. The evaluation board is known to be working, so any problems that you have when setting up and using the software toolchain can be more easily determined. If you built your own hardware before you have the toolchain running and are not familiar with the tools, it will be difficult to determine the cause of errors – is it the board that is not working, or did I not set up the program properly?

Approach Taken in this Book

Tested circuits and software are presented in this book to help you to learn microcontroller interfacing. Although not every hardware device that can be interfaced to a microcontroller has been presented, it is hoped that the selection is varied enough to be useful and to teach the principles of interfacing so that you will be able to interface other devices on your own. For this reason, specialised components have been avoided wherever possible.

Hardware and software
Chapter 1 is concerned with getting software tools installed on your computer, getting a hardware platform to work from and compiling and loading C programs to the hardware. It also takes a look at a basic ARM microcontroller circuit and examines the microcontroller selected for use with this book.

Start interfacing and programming
The chapters that follow present interfacing various hardware devices to microcontrollers and software programs that operate the interfaced hardware.

The principles of operation of the hardware devices are included so that you will understand what the software must do. This enables you to convert the software or write your own software to operate the hardware interfaced to any microcontroller.

Operation of the software is also explained, but where it is lacking in detail, you are expected to refer to the example programs that can be downloaded from the support page for this book on the Elektor website. Where reference is made to various microcontroller register and bit fields within these registers, you will need to refer to the microcontroller datasheet as it was considered unnecessary to reproduce these registers in this book.

More advanced microcontrollers
The microcontroller used throughout this book is a fast and powerful ARM microcontroller and a great step up from 8-bit microcontrollers. But what do you do if you need more processing power, I/O pins or more memory? The final chapter in this book presents additional ARM microcontrollers that will solve these problems including some of the latest offerings from Atmel, a major manufacture of such devices (see **www.atmel.com**).

Download required
Example programs and datasheets that this book refers to must be downloaded from the Elektor website. Inside the zipped file that you download, you will find a webpage (index.html) that contains links to all of the websites and datasheets referenced in the text for easy access to these resources.

Good luck and enjoy interfacing with ARM microcontrollers!

1. Software and Hardware Setup

Interfacing any device to a microcontroller is completely useless unless we have software to control the device. This chapter will show you how to set up GNU C software tools for programming ARM microcontrollers and gives some suggestions on how to prototype hardware that is used in this book.

1.1 Installing the YAGARTO GNU ARM Toolchain

YAGARTO (**Y**et **A**nother **GNU** **ARM** **To**olchain) can be downloaded from **www.yagarto.de** and consists of three main parts:

1) **YAGARTO tools** – containing the 'make' program, among others.
2) The **GNU ARM toolchain** itself containing compiler, linker assembler, gdb debugger and more.
3) **Eclipse IDE** – Integrated Development Environment that can be used to edit source code and run YAGARTO. Debugging can also be done from within Eclipse.

Links to these packages can be found on the front page of the YAGARTO website under the heading "**Download**". Detailed download and installation instructions are available on the YAGARTO website (click the "**HOW TO**" link), the steps are summarised below. Note that these steps were accurate at the time of writing, however it is possible that the website has been modified and/or updated and that the version of the software packages has changed. If new versions of the software packages are available, this will be seen in a change in the version number (if included) and date in the file names of these packages.

1.1.1 Installing YAGARTO Tools

Go to the YAGARTO home page and scroll down to the "**Download**" link. Clicking the "**YAGARTO Tools**" link under "**Package**" will enable you to download the "**yagarto-tools-20091223-setup.exe**" file. Run this file in order to install the tools.

1.1.2 Installing the GNU ARM Toolchain

Clicking the "**YAGARTO GNU ARM toolchain**" link will open a page at **sourceforge.net**. Click the "**Download Now!**" button to download the package. Run the downloaded file to install the toolchain. The name of the file downloaded and used with this book was **yagarto-bu-2.20_gcc-4.4.2-c-c++_nl-1.18.0_gdb-7.0.1_20091223.exe**.

1.1.3 Installing the Eclipse IDE

Clicking the "**Integrated Development Environment**" link opens up a page on the YAGARTO website that explains how to install the Eclipse IDE. The steps are:

1) Check if the Java Runtime Environment (JRE) is installed on your PC. If it is not, download and install it (**java.sun.com/javase/downloads/index.jsp**) Downloaded file was **jre-6u18-windows-i586.exe**.
2) Go to the Eclipse website (**http://eclipse.org/**) and click the "**Downloads**" link to go to **www.eclipse.org/downloads/** and click the "**Eclipse IDE for C/C++ Developers**" to download the Eclipse IDE. Downloaded file was **eclipse-cpp-galileo-SR2-win32.zip**.
3) Copy the **eclipse** directory out of the downloaded zip file and put it in the root directory of your **C:/** drive.
4) Create a shortcut to the **eclipse.exe** file in this directory to be able to start Eclipse conveniently – e.g. create a shortcut on your desktop.
5) Follow the YAGARTO tutorial on Eclipse when running Eclipse for the first time. (Basic steps are to select a workspace and then click the "Workbench" icon)

The tutorial on the YAGARTO website contains information on installing the "Zylin CDT plugin". The plugin provides embedded debug functionality if you are going to use a debug tool such as a wiggler or SAM-ICE JTAG emulator (see next section). The plugin is installed from within Eclipse, so follow the YAGARTO webpage instructions if you are going to use the debugger. Information is also provided on running Eclipse for the first time and opening and compiling a project, or read further for information on compiling the example programs and starting new projects in this chapter.

1.2 Hardware Setup

1.2.1 Recommended Hardware

The AT91SAM7S256 is the chosen microcontroller for this book and the AT91SAM7S-EK evaluation board can be used as a standard hardware platform. Any other board with an AT91SAM7 microcontroller should also work. To load programs into the memory of this microcontroller, the board on which it is soldered must have either a JTAG connection (20 pin IDC is standard), or the USB port must be available, or the serial DBGU port must be wired up as an RS-232 port. All three of these options are available on the AT91SAM7S-EK.

To program this microcontroller via the JTAG port, an extra hardware device must be purchased: the AT91SAM-ICE emulator (described below). To program the microcontroller via the USB port, the microcontroller must have an 18.432MHz crystal as its main clock. **Figure 1.1** shows the AT91SAM7S-EK board with SAM-ICE attached.

The **AT91SAM7S-EK** page on the Atmel website is at:
www.atmel.com/dyn/products/tools_card.asp?tool_id=3784
Information on the **AT91SAM-ICE JTAG Emulator** is available at:
www.atmel.com/dyn/products/tools_card.asp?tool_id=3892

1.2.2 Using Different ARM Microcontrollers

It is possible to use other ARM7 microcontrollers from Atmel with minimal changes to the software examples. ARM microcontrollers from other manufacturers could be used with the same software toolchain, but the example programs will have to be changed wherever hardware is accessed in the programs. The assembly language start-up file that initialises the microcontroller will need to be changed too. The linker command file will need to be set up for the memory map of the different microcontroller.

Be aware that:

- Most ARM microcontrollers operate with an I/O voltage of 3.3V, including the AT91SAM7S. This means that the I/O pins of the microcontroller will drive an interfaced circuit with a maximum of 3.3V and will be able to read an input voltage on its pins of up to 3.3V. If you choose a different microcontroller, it may operate from a different I/O voltage e.g. 5V.

- The Atmel SAM-ICE debugging tool described in this chapter only works with Atmel ARM microcontrollers. The more expensive J-link ICE is an alternative that can be used with a whole range of different ARM microcontrollers from different manufacturers.

1.2.3 Using Microcontrollers with Different Processor Cores

If a microcontroller with a different processor core architecture (e.g. PowerPC, x86) is to be used, the external circuits described in this book that are interfaced to the microcontroller will work, just remember to check the I/O voltage. Modifications would also need to be made to the example programs wherever hardware is accessed – this is assuming that the C programming language is being used. A software toolchain specific to the architecture of the microcontroller chosen would also need to be used.

1.2.4 ICE JTAG Emulator for Programming and Debugging

If the AT91SAM7S, or in fact any Atmel ARM microcontroller is used, the low cost AT91SAM-ICE JTAG emulator can be used to load programs into the memory of the microcontroller as well as be used to single-step debug the microcontroller. The JTAG emulator plugs into the PC via a USB connection and into the microcontroller via a 20 pin IDC connector.

I have tried various 'wiggler' clones and other USB devices that allow you to single step through programs, but have found the SAM-ICE to be easy to use and reliable with no setup hassles. The SAM-ICE is made by Segger – the software used with the SAM-ICE can be downloaded from the Segger website (**www.segger.com/cms/jlink-software.html**). You can download the latest version of the software at this link. The SAM-ICE uses the same software as the **Segger J-Link emulator**. The main difference between the SAM-ICE and the J-Link is that the SAM-ICE can only be used with Atmel ARM microcontrollers, whereas the J-Link can be used with almost any manufacturers ARM microcontroller. The J-Link is, however, a lot more expensive than the SAM-ICE. For **non-commercial/educational use** the J-Link is available at a reduced price. Details are here:
www.segger.com/cms/j-link-arm-for-non-commercial-use.html
OR
www.segger.com/cms/j-link-edu.html

1.2.5 Interfacing Suggestions

Of course it is possible to design your own ARM microcontroller PCB from scratch, but when experimenting or prototyping it is easier and cheaper to use a pre-made board such as the AT91SAM7-EK. Wires can be connected via the 44 pin header to a solderless breadboard and the circuits built and tested on the breadboard. For a more permanent arrangement, vero-board can be used.

The circuits in this book were all tested using a simple AT91SAM7S256 board built and set up much like the AT91SAM7S-EK, but without the switches, LEDs, extra serial port and buffer on the ADC channels. When interfacing surface mount components to the microcontroller during prototyping, leads were soldered to the surface mount components so that they could be plugged into the breadboard. Higher power components were soldered onto vero-board for testing. **Figure 1.2** shows the setup used for testing circuits. The figure shows the AT91SAM7S256 board interfaced to MOFETs for driving a stepper motor and an LCD display interfaced to I/O pins. It looks messy, but worked well! Although perhaps difficult to see in the photo, there are surface mount opto-couplers with leads soldered to them inserted into the breadboard.

If you are using the AT91SAM7S-EK, be sure to look at the circuit diagram to see which pins are already connected to on-board hardware. For example, port pins PA0 to PA4 are connected to on-board LEDs. If you want to interface these pins to external hardware, you may want to break the tracks that connect the pins to the LEDs. Pads are intentionally provided on the board to break these tracks and then re-join them with a blob of solder. See **figure 1.3**.

The AT91SAM7S-EK can be powered via the USB port on a PC or an external power supply. You may prefer to use an external supply when doing experimental work rather than risk damaging the USB port on your PC. Also it may happen that your PC, laptop or netbook might not have sufficient USB drive capability to maintain the necessary voltages for the board. A diode that connects the supply from the USB port to the board can be removed as a safety precaution – see the circuit diagram.

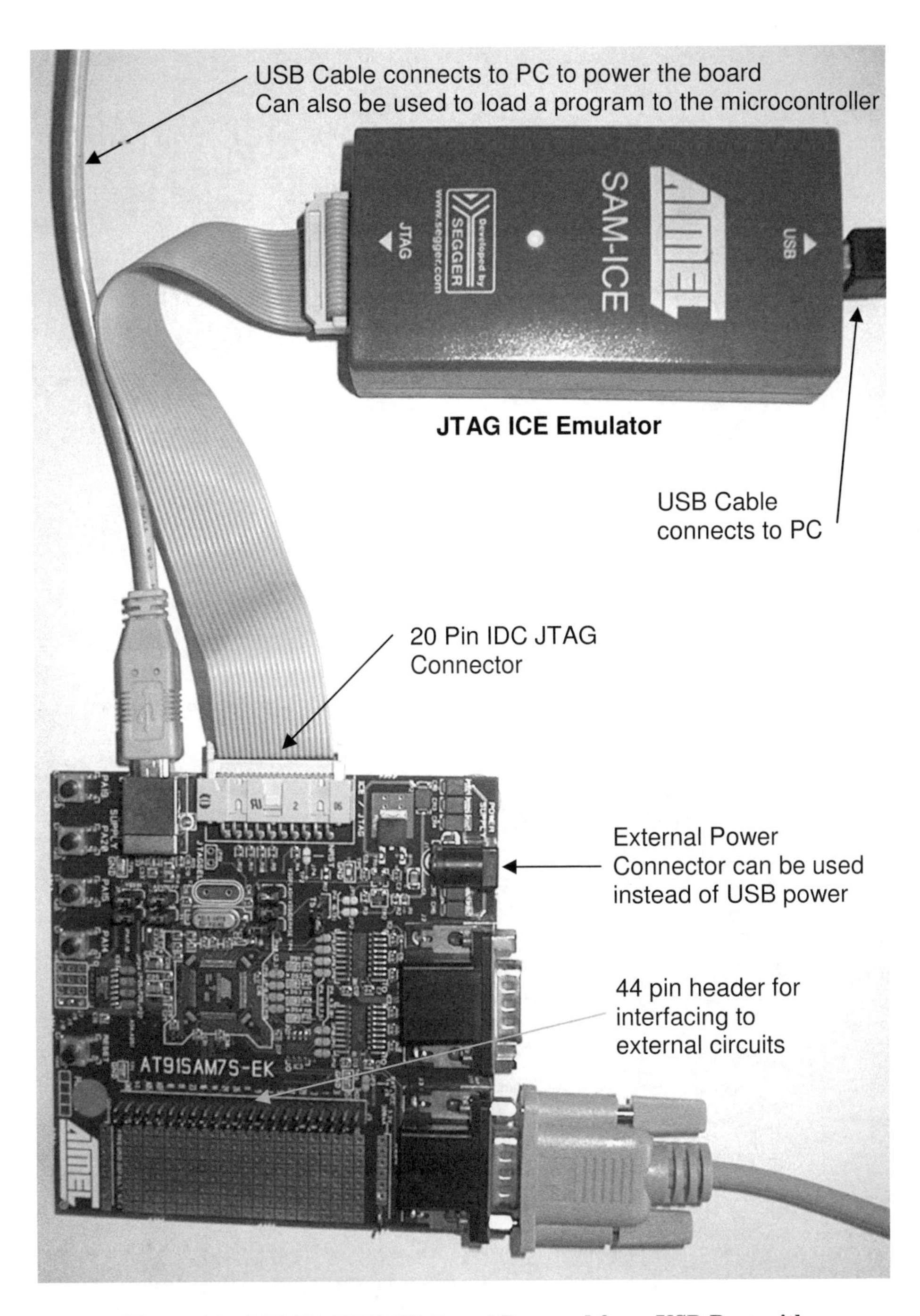

Figure 1.1: AT91SAM7S-EK Board Powered from USB Port with AT91SAM-ICE JTAG Emulator

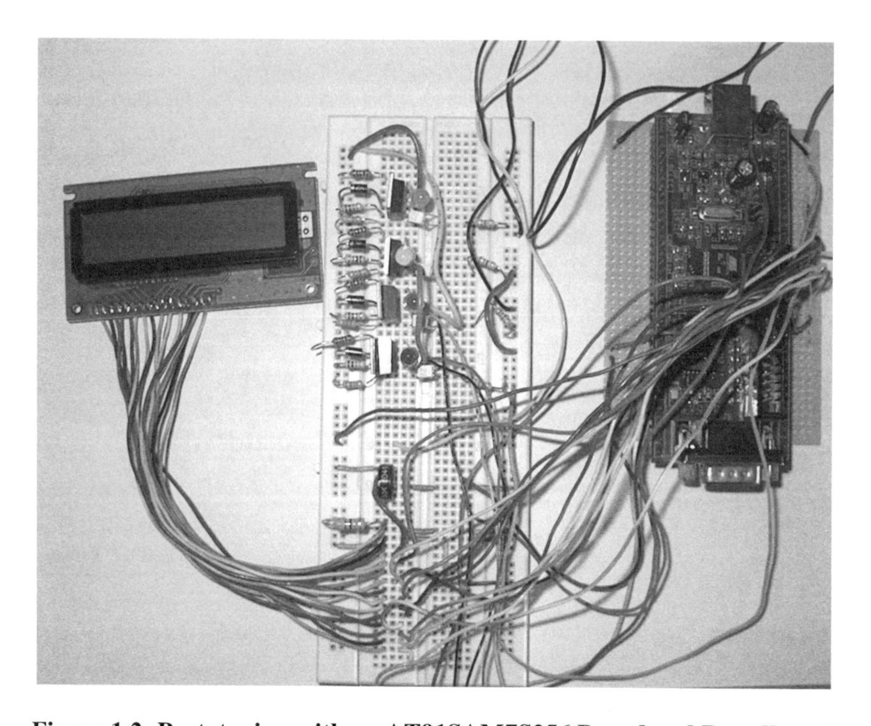

Figure 1.2: Prototyping with an AT91SAM7S256 Board and Breadboard

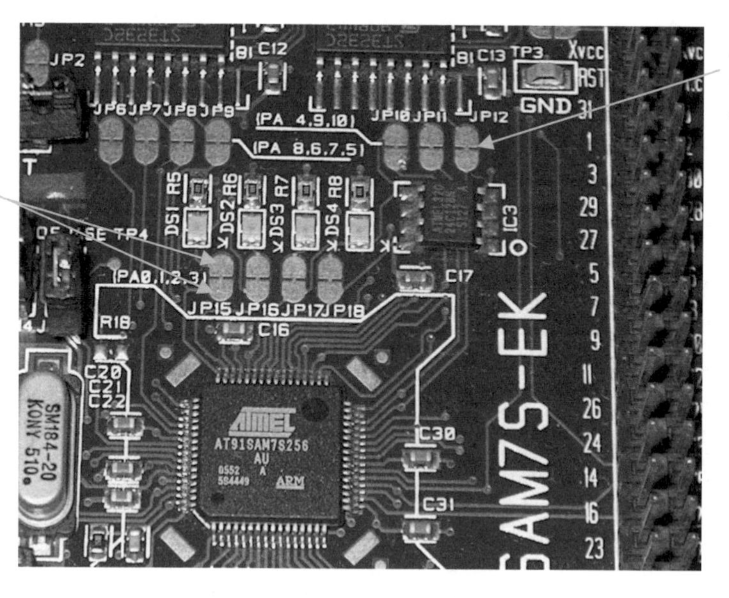

Figure 1.3: Hardware Devices on the AT91SAM7S-EK can be Connected and Disconnected with Solder Pads

1.3 Loading a Program to the Microcontroller

After a program has been compiled and linked and a binary output file produced (described later), the binary file must be loaded into the Flash memory of the microcontroller so that it can be run. A utility program from Atmel called SAM-BA that runs on a PC is used to load the binary file into the microcontroller.

SAM-BA loads programs into AT91SAM7S microcontrollers via three different methods:

1) Via a direct cable connection from the PC USB port to the microcontrollers USB port.
2) Via a direct cable connection between the PC RS-232 port and the microcontroller DBGU RS-232 port.
3) Via a SAM-ICE JTAG emulator connected to the JTAG port of the microcontroller.

To use a direct USB or RS-232 connection, the SAM-BA bootloader program that is on the AT91SAM7S microcontroller must be loaded to Flash memory so that it runs as soon as the microcontroller is initially powered up (a so-called ‘cold-start’). This is done by pulling the TST pin of the microcontroller high, switching power to the board on for a few seconds, switching the power off and then pulling the TST pin low and powering the board up again.

To load programs using the SAM-ICE JTAG emulator and the SAM-BA PC program to the microcontroller, the software for the SAM-ICE must be loaded on the PC – see section **1.5 Debugging** for information on loading the emulator software. It is not necessary to load the on-chip SAM-BA bootloader program to the microcontroller’s Flash memory when using the SAM-ICE.

Installing SAM-BA on the PC

The SAM-BA PC program can be downloaded from the Atmel website (**www.atmel.com**). On the website, click **Products** and then **ARM-based solutions** under **Microcontrollers**. Click **Tools & Software** and then **Programmer**. Click the **AT91 In-system Programmer (ISP)** link. Download the latest version of **AT91-ISP** and run it to install SAM-BA. The downloaded file was **Install AT91-ISP v1.13.exe**.

Hardware requirements when using SAM-BA with the USB port
To use SAM-BA to program the AT91SAM7S256, the hardware must be set up as follows (the AT91SAM7S-EK board is set up like this):

- The USB lines from the microcontroller must be connected to a USB connector.
- The microcontroller must be using an 18.432MHz crystal for its oscillator.
- The TST pin must be connected to a jumper so that it can be pulled high to load the on-chip SAM-BA bootloader to Flash memory.
- Port pin PA16 must be dedicated to the USB pull-up resistor.
- Port pins PA0 to PA3 must be pulled high when the TST pin is pulled high to load the SAM-BA bootloader to Flash.

I have found that SAM-BA will work with a permanent 1k5 pull-up resistor connected to the USB DP line, hence PA16 is not needed to be dedicated to the pull-up. See chapter 8 “Serial Ports” for more details on USB (**8.4 USB**).

It is also a good idea to have a jumper connected to the ERASE pin of the microcontroller should you set one of the lock bits in the microcontroller that will prevent SAM-BA from being loaded to Flash memory again. Powering up the board with the ERASE pin pulled high will erase all Flash and lock bits.

1.4 Using the DBGU Serial Port

Some of the example programs in this book use the DBGU serial port to send and receive information between the embedded system and the PC. One example is when interfacing a keypad to the microcontroller, the ASCII value of the key pressed is sent out of the serial port so that it can be displayed on a terminal emulator program running on the PC. This is the simplest way of testing that the keypad interfacing program works, as not every embedded system will have an LCD or other display.

On the AT91SAM7S-EK board, the DBGU serial port is set up as an RS-232 port and can be connected to the PC serial port via a serial cable with the transmit and receive lines crossed (DBGU Tx pin connected to PC Rx pin, DBGU Rx pin connected to PC Tx pin). A terminal emulator such as Hyperterminal on Windows can be run in order to communicate with the embedded system.

If your PC does not have an RS-232 serial port, you can buy a low cost USB to serial RS-232 port adapter. Windows Vista no longer comes with Hyperterminal, so an alternate terminal emulator will need to be downloaded and installed on the PC.

Some alternate terminal emulators are:
Bray Terminal from **http://sites.google.com/site/braypp/terminal**
RealTerm from **http://realterm.sourceforge.net/**

The DBGU port settings and therefore the terminal emulator settings used throughout the book are the following:

Baud rate (Bits per second):	**115200**
Data bits:	**8**
Parity:	**None**
Stop bits:	**1**
Flow control:	**None**

RS-232 cable
The cable must connect the Tx pin of the AT91SAM7S to the Rx pin on the PC and the Rx pin of the AT91SAM7S to the Tx pin of the PC RS-232 port. On the AT91SAM7S-EK a cable must be used with two female DB-9 connectors with pin 2 of the first connected to pin 3 of the second and pin 3 of the first connected to pin 2 of the second. Pin 5 is the common ground pin and must also be connected between the two.

Running the terminal emulator software
Whatever terminal emulator you decide to use, the basic steps for connecting to the microcontroller serial port are:

1) If your PC does not have an RS-232 port, plug a USB to RS-232 adapter into the PC (it may come with software that must be installed).
2) Plug the RS-232 cable between the PC and the microcontroller board.
3) Start the terminal emulator software.
4) Select the COM port to connect to – e.g. COM3.
5) Enter the communication parameters (baud rate, etc.) as described above.
6) Click the connect button.
7) Power on the microcontroller board.

1.5 Debugging

Debugging refers to searching for errors in a program and correcting them. The compiler and linker will notify you of some of the errors that may be present in a program at compile time. These are usually syntax errors or errors to do with functions or objects that are missing during linking.

Some program errors (or bugs) will only be discovered when the program is running. An example of one of these run-time errors is a hardware device that is not initialised properly or a program that crashes at some stage. The compiler would not be able to check for bugs like these.

The simplest form of run-time debugging is for the programmer to use an LED, port pin or serial port as a way of indicating what is happening in the program. LEDs can be switched on at certain places in a program to see how far the program runs before it crashes. A port pin on the microcontroller can be connected to an oscilloscope and used in a similar fashion. Diagnostic information can be sent via the serial port to a terminal emulator program running on the PC.

Another method is to use an in-circuit emulator (ICE) such as the Atmel SAM-ICE JTAG emulator already mentioned in this chapter. The ICE is connected to the PC via a USB cable and the JTAG connector (20-pin IDC connector is standard for ARM microcontrollers) is plugged into the microcontroller board. When software is run on the PC, it allows the JTAG ICE to single step through the C source code of the program and the values of registers and variables can be examined, breakpoints can be inserted into the program so that it will stop automatically when reached by the program.

Software to install for using the Atmel SAM-ICE JTAG emulator
A piece of software called the **Zylin Embedded CDT plugin** must be installed in Eclipse which adds debug support for Elipse. The YAGARTO website provides instructions on how to install this plugin. The installation is done from within Eclipse.

Segger manufacture the AT91SAM-ICE and provide the software for use with this device. The newest software for use with the ICE can be downloaded from the Segger website (**www.segger.com**).

To download the latest software:

1) On the Segger website, click the **Downloads** link.
2) In the **Products** drop down list, select **J-Link ARM**.
3) Click the download link that appears.
4) The **J-Link downloads** page will open (**www.segger.com/cms/jlink-software.html**).
5) Click the **Software and documentation pack** link and download the zip file. Download file was **Setup_JLinkARM_V412.zip**.

The zipped file contains an executable installation program that will install the software used with the SAM-ICE.

When using the SAM-ICE for debugging, the **J-Link GDB Server** program that is part of the Segger software that was installed must be started before debugging.

The YAGARTO website and a tutorial by James P. Lynch are the best sources of information on using the SAM-ICE for debugging with Eclipse. See the references section at the end of this chapter for information on the tutorial.

Programs can be run from and debugged in both Flash and SRAM memory of the AT91SAM7S microcontrollers. I find that loading the entire program into SRAM is more convenient than programming the Flash when doing experimental work. I have found that it is best to erase the Flash before running programs from RAM. If there is a program in Flash, it will start running as soon as power is applied to the microcontroller. The program that runs will set up hardware so that the microcontroller is not in the same state as it would be if it had just been reset. The Flash memory can be erased using the SAM-BA PC program.

When using the SAM-ICE to debug a program in Flash memory, a maximum of two breakpoints can be set. When debugging in SRAM, any number of breakpoints can be set. The AT91SAM7S has 64k of SRAM, if programs get bigger than this, there is no option but to debug in Flash only, although parts of a program can be compiled separately and tested in SRAM.

1.6 Template Files and Example Programs

1.6.1 Template Files and Example Programs Used in this Book

Template files have been provided with this book and can be downloaded from the Elektor website. The template files allow you to start a new project by simply copying the template files to a new directory, renaming the project in the make file, opening the project in Eclipse and then start writing a program in C. The template files have been set up for the AT91SAM7S256 and contain all of the common files that you will need for every project such as the assembly language start-up file, the make file, linker command file, etc.

The example files for each project are based on the template files – each project was started from a set of template files.

Four sets of template files have been provided:

`Flash_interrupt`	Code is set up to run from Flash memory and contains interrupt handling code in the assembly language start-up file to handle nested interrupts.
`Flash_template`	Code is set up to run from Flash memory and contains no interrupt handling code.
`RAM_interrupt`	Code is set up to run from SRAM memory and contains interrupt handling code in the assembly language start-up file to handle nested interrupts.
`RAM_template`	Code is set up to run from SRAM memory and contains no interrupt handling code.

The above names are the names of the directories (folders) in which the code is stored. Any of the projects can be started with the interrupt versions of the templates. Having the non-interrupt versions as well allows you to compare the assembly language file **`startup.s`** in both versions to see what has been added to support nested interrupts should you be interested.

The templates are structured as follows (contents of the template directories):

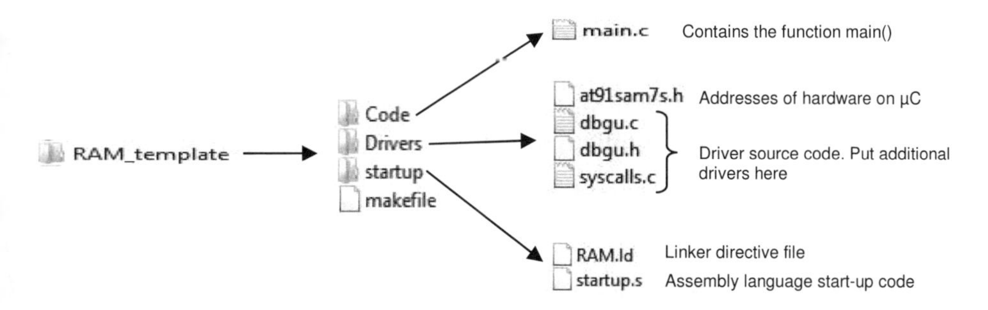

The **Code** directory contains the file **main.c** which contains the entry point of the C source code – the function **main()**. Additional source code files can be added to this directory. Whenever a new source code file is added, the make file must be updated so that the new code will be compiled and linked.

The **Drivers** directory contains the header file **at91sam7s.h** which contains the addresses of registers of on-chip hardware devices so that these registers can be programmed. **dbgu.c** and **dbgu.h** contain code for initialising the DBGU serial port, reading data from it and writing data to it. This is primarily used to display information on a terminal emulator. **syscalls.c** is necessary because of the way that the library files of YAGARTO have been compiled. It is only necessary when using functions such as **sprintf()**, so can be removed if these functions are not used.

The **startup** directory contains **RAM.ld** (or **Flash.ld** in the Flash templates) which tells the linker where memory is located on the microcontroller. The memory sizes are set up in this file for the AT91SAM7S256 and will have to be changed here should one of the other AT91SAM7S microcontroller family members be used. The file **startup.s** contains the assembly language start-up code which initialises stacks and the oscillator and a few other things before calling **main()** in the C code. If the size of the stack needs to be adjusted, it is done so in this file.

makefile contains the build instructions for the project. If source files are to be added or removed from the project, the instructions for compiling and linking these files must be added or removed from the make file.

When developing code for this book, most of the projects were developed from the RAM templates. The programs were loaded into SRAM from within Eclipse using the SAM-ICE.

Starting a new project from the template files

1. Start by copying one of the template directories to wherever you want to create your project.
2. Rename the directory to the name of your project. E.g. `temperature_logger`.
3. In Eclipse, click **File → New → C Project**
4. In the **C Project** dialogue box, uncheck **Use default location** and then use the **Browse...** button to navigate to the new project directory that you made, click it and then click the **OK** button.
5. Back in the **C Project** dialogue box, name the project in the **Project name:** field. You can give it the same name as the project directory.
6. Click the **Finish** button.
7. The new project will now appear in the **Project Explorer** pane. Click the arrow symbol to the left of the project name in this pane to expand it. Double click the **makefile** to open it in Eclipse.
8. At the top of the **makefile**, change **proj_name** to the name of your new project.
9. Click the "Hammer" icon, or right click on the project name in the **Project Explorer** pane and click **Build Project** to build the project. This should produce .elf and .bin output files.

Debugging a RAM project using the SAM-ICE

If the project was created from one of the **RAM** templates, then use the following procedure to load the program to SRAM memory using the Atmel SAM-ICE. It is assumed that you have installed the SAM-ICE software from Segger and that you have installed the plugin for Eclipse as described on the YAGARTO webpage. It is also assumed that you have compiled the project and followed the steps on the YAGARTO HOWTO webpage that describes selecting "**GNU Elf Parser**" under **Project → Properties**.

If you haven't done these steps, go to the YAGARTO website and do them now. You can ignore most of the other steps on the web page such as importing the project.

1. Plug the SAM-ICE into the PC's USB port.
2. Plug the 20 pin IDE connector of the SAM-ICE into the JTAG header on the board.
3. Start the **J-Link GDB Server** program which was installed with the Segger software.
4. In Eclipse in the **Project Explorer** pane, left click the project name so that it is highlighted.
5. Now click the down arrow next to the **Debug** icon in Eclipse (The debug icon looks like a small green bug).
6. Click **Debug Configurations…** on the menu that drops down.
7. Right click **Zylin Embedded debug (Native)** and then **New** on the menu that pops up.
8. The correct names will be added to the fields at the right of the dialog box only if the "GNU Elf Parser" step was setup as mentioned above. If it was not, you will have to manually search for the .elf file. It will also not work if you didn't start this process by selecting the project name as in step 4 above.
9. Click the **Debugger** tab in the currently open dialog box and make sure that the **GDB debugger:** field contains **arm-elf-gdb** and that **GDB command file:** is empty.
10. Click the **Commands** tab and copy and paste the contents of the **SRAM.txt** file to the **'Initialize' commands** box. A link to the **SRAM.txt** file is provided on the webpage (**index.html**) in the zip file downloaded from the Elektor website.
11. Click the **Apply** button at the bottom of the dialog box.
12. Switch the power to the embedded system on.
13. Click the **Debug** button in the dialog box.
14. Click the **Resume** icon (Green Arrow) to run the program.
15. The next time that you want to debug the program, just power up the board, start the J-Link GDB server, click the down arrow next to the debug icon and select the name of the project.

Don't forget to start the J-Link GDB Server program each time that you want to debug. In the Segger J-Link GDB Server application window, the "Initial JTAG speed" can be set to "Adaptive". "Little endian" must also be selected. ARM processors can be operated as big or little endian processors, but in the Atmel microcontrollers, they have been permanently set to little endian. The tutorial by James P. Lynch referenced at the end of this chapter will describe how to use breakpoints and other debugging functions.

Debugging a Flash project using the SAM-ICE
This is similar to debugging a RAM project, but the program must first be loaded to Flash memory.

1. Plug the SAM-ICE into the PC's USB port.
2. Plug the 20 pin IDE connector of the SAM-ICE into the JTAG header on the board.
3. Apply power to the board.
4. Start the SAM-BA application on the PC.
5. A dialog box pops up when SAM-BA is started, the connection must be set to "**\jlink\ARM0**" to select the SAM-ICE and board to "**at91sam7s256-ek**" if your board contains an AT91SAM7S256. Click the connect button.
6. In the SAM-BA application, the "Flash" tab will be selected as default. Use the "open folder" icon next to the "Send File" button to navigate to the binary file that you will download to the microcontroller. Select it and click the "Open" button in the dialog box.
7. Click the "Send File" button.
8. When you are asked about locking regions, click "No".
9. Close SAM-BA and power the microcontroller off.
10. Start the **J-Link GDB Server** program which was installed with the Segger software.
11. In Eclipse in the **Project Explorer** pane, left click the project name so that it is highlighted.
12. Now click the down arrow next to the **Debug** icon in Eclipse (The debug icon looks like a small green bug).
13. Click **Debug Configurations…** on the menu that drops down.
14. Right click **Zylin Embedded debug (Native)** and then **New** on the menu that pops up.
15. The correct names will be added to the fields at the right of the dialog box only if the "GNU Elf Parser" step was setup as mentioned above. If it was not, you will have to manually search for the .elf file. It will also not work if you didn't start this process by selecting the project name as in step 11 above.

16. Click the **Debugger** tab in the currently open dialog box and make sure that the **GDB debugger:** field contains **arm-elf-gdb** and that **GDB command file:** is empty.
17. Click the **Commands** tab and copy and paste the contents of the **Flash.txt** file to the **'Initialize' commands** box. A link to the **Flash.txt** file is provided on the webpage (**index.html**) in the zip file downloaded from the Elektor website.
18. Click the **Apply** button at the bottom of the dialog box.
19. Switch the power to the embedded system on.
20. Click the **Debug** button in the dialog box.
21. Click the **Resume** icon (Green Arrow) to run the program or click one of the step buttons to single step.
22. The next time that you want to debug the program, just power up the board, start the J-Link GDB server, click the down arrow next to the debug icon and select the name of the project. Always ensure that you have the current version of the program that you are going to debug loaded into Flash memory.

Converting a RAM project to Flash

You may have started a project that runs in the microcontroller's SRAM and decide that you want it to run from Flash memory either because it has grown too big for SRAM or it is a completed project and you wish the program to start running at power-up. To convert the project, you need to do the following:

1. Make a copy of the RAM version of the project and rename it – e.g. with "Flash" in the name.
2. Replace the make file with the make file from the Flash template.
3. Modify the make file to add or remove any files to compile and link with the project that you may have added to or removed from the original RAM project. Also change the name of the project in the make file.
4. Copy **Flash.ld** from the template files to the **startup** directory. This will replace **RAM.ld**.
5. **startup.s** in the **startup** directory may have to be replaced if you are changing from a project that does not use interrupts to one that does.
6. Create a new project in Eclipse from the new project directory that you have created.
7. Build the project.

Changing the linker control file to support other microcontroller memory maps

If you are using any other AT91SAM7S besides the AT91SAM7S256, you will need to change the size of memory in the linker control file. This file is found in the **startup** directory of the templates and example projects and will be called either **RAM.ld** or **Flash.ld**.

In **Flash.ld**:

```
MEMORY
{
      CODE (rx) : ORIGIN = 0x00100000, LENGTH = 0x00040000
      DATA (rw) : ORIGIN = 0x00200000, LENGTH = 0x00010000
}
```

The LENGTH parameter in the CODE line must be changed for different Flash sizes and the LENGTH parameter in the DATA line must change for different SRAM sizes.

Examples:

AT91SAM7S512 512k Flash – LENGTH = 0x80000 64k SRAM – 0x10000
AT91SAM7S256 256k Flash – LENGTH = 0x40000 64k SRAM – 0x10000
AT91SAM7S128 128k Flash – LENGTH = 0x20000 32k SRAM – 0x8000
AT81SAM7S64 64k Flash – LENGTH = 0x10000 16k SRAM – 0x4000

Whenever the size of the SRAM is changed in a linker directive file, the address of the top of the stack must be changed in the **startup.s** file:

```
/* top of the stack in RAM */
    .equ    TOP_STACK,      0x0020FFFC
```

The top of the stack is set to the last word address in RAM. This address is the start address of the SRAM plus the length of the SRAM minus 4.

1.6.2 Using the Atmel Example Programs

Atmel also have example programs for use with their evaluation kits. For the AT91SAM7S-EK, example programs are contained in the **at91sam7s-ek.zip** file that can be downloaded by clicking the **AT91SAM7S-EK Software Package for IAR 5.2, Keil and GNU** link on the AT91SAM7S-EK page at the Atmel website.

The zip file unpacks to the **at91sam7s-ek** directory. One of the files in this directory, **index.html**, can be opened in a web browser. It contains information on the example programs as well as links to documentation and compiler vendor websites. The **packages** directory contains the zipped source code for the example programs. Be sure to choose the packages that end with 'gnu' if you are using the GNU toolchain.

The example programs contain the binary output files that are generated from building the project. The make file is set up by default to use the Sourcery G++ compiler (**www.codesourcery.com** – the Sourcery G++ Lite for ARM toolchain can be downloaded for free). If you are using YAGARTO, you will need to change the following in the Atmel examples make file:

```
# Tool suffix when cross-compiling
CROSS_COMPILE = arm-none-eabi-
```

Must change to:

```
CROSS_COMPILE = arm-elf-
```

Also to look out for in the make file is if the board and chip are set up correctly. The Atmel software examples are fairly universal and so can be used across several of its microcontrollers, so sometimes the example programs will have a different board and chip set up as the default to what you are using. You may have to change, for example:

```
CHIP  = at91sam7se512
BOARD = at91sam7se-ek
```

To:

```
CHIP  = at91sam7s256
BOARD = at91sam7s-ek
```

To build the project, change into the project subdirectory that contains the make file from a command prompt in Windows. This subdirectory is named after the project. Type 'make clean' to delete all object and binary files and 'make' to build the project.

1.7 Microcontroller Hardware

This section talks about the microcontroller and microcontroller family chosen as the main microcontroller used in this book, namely the AT91SAM7S256 and the AT91SAM7S microcontroller family.

1.7.1 AT91SAM7S Packaging

AT91SAM7S microcontrollers are available in two different types of physical packages namely **LQFP** (Low-profile Quad Flat Package – see **figure 1.4**) and **QFN** (Quad Flat No-leads). The pin numbering on each type of package is exactly the same; the QFN package however, does require the bottom pad to be connected to ground. The LQFP package is easier for hobbyists to solder to a board and is the same package type that is used on the AT91SAM7S-EK board.

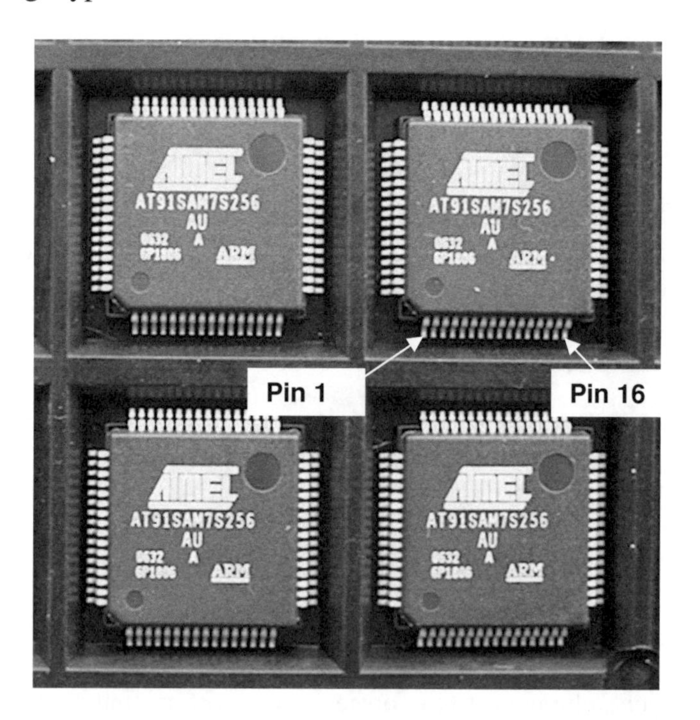

Figure 1.4: AT91SAM7S256 Microcontrollers in 64-pin LQFP Packages

The AT91SAM7S family of microcontrollers were originally available in 64 pin packages, but additional members were added to this family that are packaged in 48 pin packages. The 48 pin packaged variety (AT91SAM7S32 and AT91SAM7S16) have reduced functionality and I/O pins.

1.7.2 AT91SAM7S Hardware

AT9SAM7S microcontrollers operate from a 3.3V I/O voltage, but the core operates from 1.8V. A regulator is built into the microcontroller that will derive the 1.8V from the 3.3V, so no external 1.8V regulator is required.

These microcontrollers have 32 general purpose I/O lines (except for the 48 pin variety that have 21 I/O pins) that can be individually configured as inputs, outputs or connected to one of the internal peripheral devices. **Figure 1.5** shows a block diagram of the 64 pin AT91SAM7S microcontrollers.

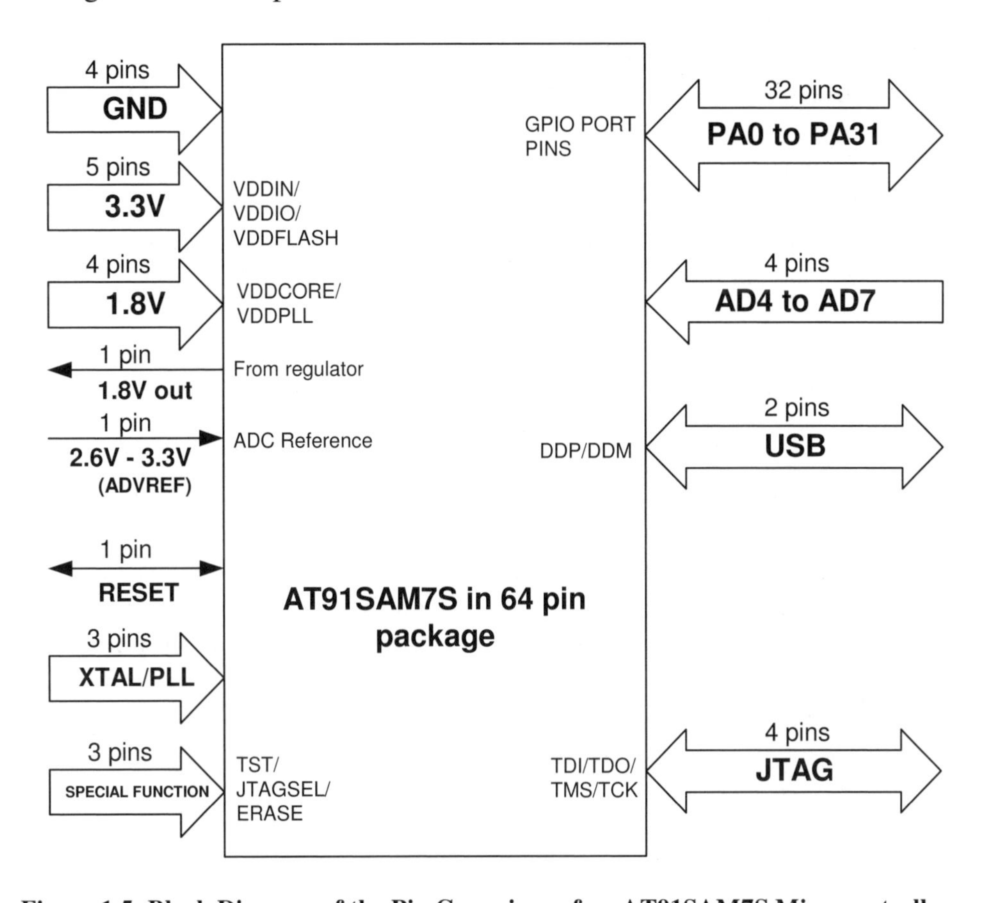

Figure 1.5: Block Diagram of the Pin Groupings of an AT91SAM7S Microcontroller

Internal peripheral devices available on the AT91SAM7S microcontrollers that can be connected to the I/O pins (multiplexed) are:

- **Pulse width modulator (PWM)** – 4 channels: PWM0 to PWM3.
- **Timer counter (TIO)** – 3 channels: TIO0 to TIO2.
- **Two Wire Interface (TWI)** – 1 I²C compatible channel.
- **Universal Synchronous/Asynchronous Receiver Transmitter (USART)** – 2 USARTs.
- **DBGU UART** – 1 UART that forms part of the debug unit.
- **Serial Peripheral Interface (SPI)** – 1 master/slave SPI.
- **Synchronous Serial Controller (SSC)** – 1 SSC capable of interfacing to I²S devices (audio CODECs).
- **Advanced Interrupt Controller (AIC)** – 2 external interrupts (IRQ0 and IRQ1).
- **Fast Interrupt (FIQ)** – 1 external fast interrupt.
- **Parallel Input/Output Controller (PIOA)** – controls setting up of pins as inputs or outputs, connecting pins to internal peripherals, setting or clearing pins that are set up as outputs, reading the state of pins that are set up as inputs.

Other hardware devices connected to microcontroller pins but not necessarily multiplexed:

- **Universal Serial Bus (USB)** – 1 USB 2.0 full speed device port (12M bits per second).
- **Analogue to Digital converter (ADC)** – 1 8-channel 10-bit ADC. ADC pins AD4 to AD7 are dedicated pins on the microcontroller, AD0 to AD3 are multiplexed with I/O pins. ADC can run in 10-bit or 8-bit mode.

Internal hardware devices are:

- **Reset Controller (RSTC)** – provides power-on reset and brown-out (i.e. power drops below the minimum specified supply voltage) detection, so no external reset circuitry is necessary. Can also drive the reset pin to reset external devices. Can be connected to an external reset switch for user reset of microcontroller.
- **Clock Generator (CKGR)** – for producing the main clock for the microcontroller. When the microcontroller is powered up, it starts running from an internal RC slow clock. The main clock is set up in software immediately after power up.
- **Power Management Controller (PMC)** – controls the clocks to various internal peripheral devices.
- **Advanced Interrupt Controller (AIC)** – automatically vectors interrupts, can also connect to external interrupt lines as mentioned under multiplexed hardware devices.
- **Periodic Interval Timer (PIT)** – can be used for operating system's scheduler interrupt or as a general purpose timer.
- **Watchdog Timer (WDT)** – for resetting the microcontroller if software fails to run a 'watchdog kick' routine within a specified time i.e. seen as a software crash. The WDT can be disabled if not required.
- **Real-time Timer (RTT)** – runs from internal RC oscillator
- **Peripheral DMA Controllers** – 11 peripheral DMA controllers
- **Timer/Counter (TC)** – 3 channel timer counter that can be used as general purpose timers or connected to external pins for timing or counting external events.

The **48 pin versions** of the AT91SAM7S family differ from the above in the following:

- Have 1 external interrupt source
- 21 I/O lines
- 9 peripheral DMA controller channels
- No USB port
- 1 USART

Each hardware device is controlled using software by reading/writing memory mapped registers.

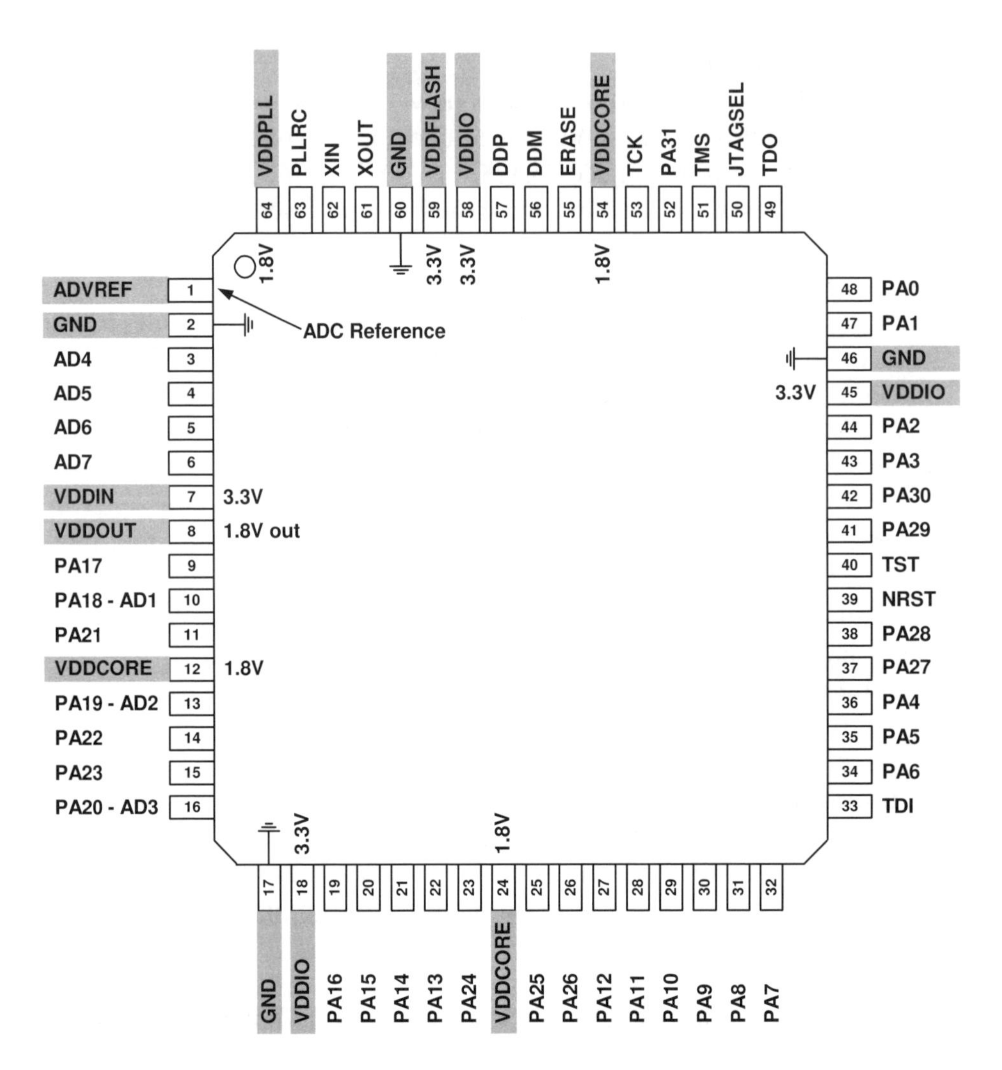

Figure 1.6: Top View of an AT91SAM7S Microcontroller in a 64 Pin Package with the Power Pins Highlighted

AT91SAM7S Pin Functions

Figure 1.6 shows the top view of an AT91SAM7S microcontroller in a 64 pin package.

This figure highlights the power supply pins of the microcontroller:

- **ADVREF** is the ADC reference voltage and can be between 2.6 and 3.3V.
- **GND** pins connect to 0V of the power supply.
- **VDDIN** is the 3.3V supply to the on-chip regulator and ADC.
- **VDDOUT** is the 1.8V from the on-chip regulator that will be used to supply the microcontroller core and PLL.
- **VDDCORE** supplies 1.8V to the ARM core.
- **VDDIO** supplies the 3.3V I/O lines and USB transceivers.
- **VDDFLASH** is the 3.3V supply to the on-chip Flash memory.
- **VDDPLL** supplies 1.8V to the oscillator and the Phase Locked Loop (PLL).

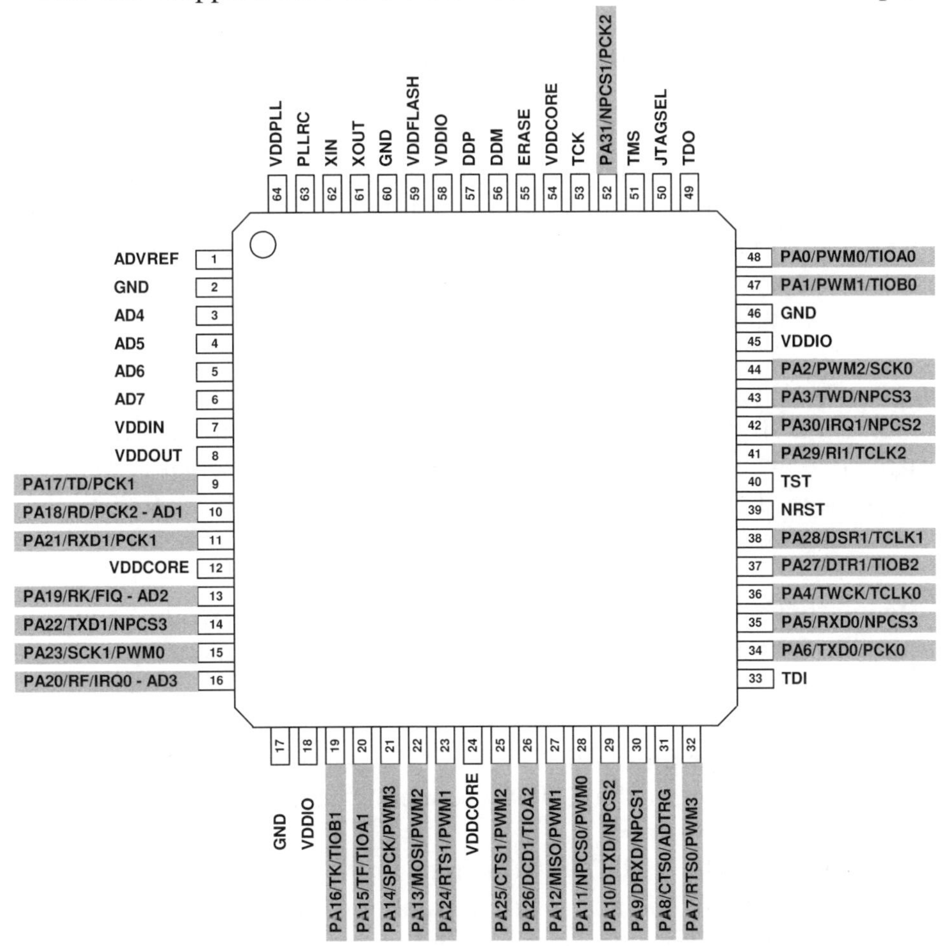

Figure 1.7: Top View of an AT91SAM7S Microcontroller in a 64 Pin Package with the Multiplexed I/O Pins Highlighted

Figure 1.7 shows the top view of an AT91SAM7S microcontroller with the multiplexed functions of the I/O pins highlighted. The 32 bit port of the AT91SAM7S is labelled AD0 to AD31. Each pin is multiplexed with two additional internal peripherals. For example pin 48 can be used as general purpose I/O pin PA0 or connected to pulse with modulator PWM0 or connected to a timer counter channel 0 (TC0) pin called TIOA0. This pin can effectively be programmed as one of four different things: a general purpose input, a general purpose output, a PWM channel or a timer counter pin.

1.7.3 Basic microcontroller circuit

Figure 1.8 shows a basic AT91SAM7S microcontroller circuit diagram. Let's examine each part of this circuit:

Power Supply: In the USB block is a 3-pin header (J4) that can select between an external 5V power supply or the 5V from the USB connection. EXT_5V on pin 3 of J4 would join to a connector on the board (not shown) that would connect to the external supply. A jumper would link the USB power or external supply to pin 2 of J4. This would in turn supply the 3.3V LDO regulator.

When you design your own board, you can use the arrangement shown or if you are building a USB device, permanently supply the board from the USB 5V. Alternatively power can be permanently supplied from an external power supply. 5V may be needed to operate other devices that may be interfaced to the microcontroller as described in the interfacing chapters in this book.

3.3V Regulator: The regulator block shows a LDO (Low Drop-Out) regulator that drops the 5V power supply to 3.3V that will supply the microcontroller and other devices on the board.

Microcontroller power pins: All 3.3V and 1.8V power pins on the microcontroller are connected to 100nF decoupling capacitors.

On-chip regulator: The VDDIN pin (microcontroller pin 7) of the regulator must be connected to a 100nF NPO capacitor in parallel with a 4.7μF X7R capacitor. The VDDOUT pin (microcontroller pin 8) of the regulator must be connected to a 1nF NPO capacitor in parallel with a 2.2μF X7R capacitor. The decoupling capacitors must be placed physically as close to the microcontroller pins as possible.

Crystal oscillator: A crystal with a value of between 3 to 20MHz can be used with the AT91SAM7S. If the USB port is to be used, only an 18.432MHz crystal can be used as shown in the schematic diagram of **figure 1.8**. This is the same crystal value used on the AT91SAM7S-EK board.

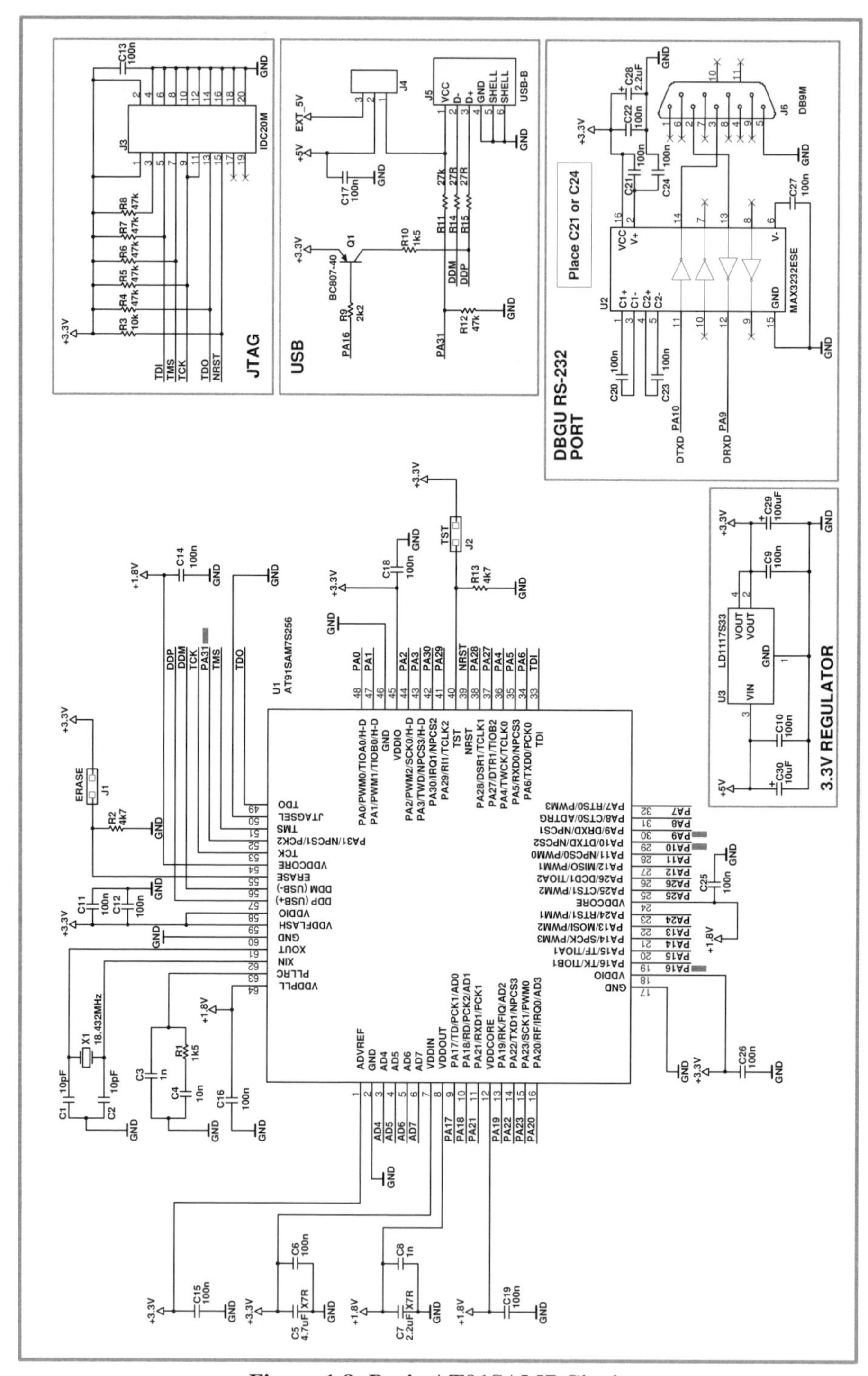

Figure 1.8: Basic AT91SAM7 Cicuit

PLL filter circuit: The PLLRC pin (microcontroller pin 63) must be connected to a second order filter as shown in the schematic diagram. The filter shown uses the correct capacitor and resistor values for the chosen frequency. These values must be re-calculated if a different clock frequency is to be used. Atmel provides an Excel spreadsheet to do these calculations.

ERASE pin: The ERASE pin is pulled to ground via a 4k7 resistor but can be pulled high via a jumper connected to J1 to erase all Flash memory and lock bits.

TST pin: A 4k7 resistor pulls the TST pin low (inactive) and a jumper (J2) enables the test pin to be pulled high. Pulling the TST pin high enables the re-loading of the SAM-BA on-chip boot loader program to Flash memory thus enabling user programs to be loaded to Flash via the USB or DBGU serial ports. When a user program is loaded to Flash memory, it overwrites the boot loader in Flash. To re-load the boot loader, the TST pin must be pulled high, power to the board switched on for 10 seconds, then off and then the TST pin must be pulled low again.

JTAGSEL pin: This pin can be pulled high for JTAG boundary scan functionality. In the schematic it is pulled permanently low to disable this functionality. Keeping this pin low enables EmbeddedICE mode to be used which allows debugging through a JTAG ICE.

JTAG connector: The schematic shows a 20 pin IDC connector (J3) used for the JTAG port. This allows a JTAG emulator such as the SAM-ICE to be directly plugged into the board to load programs to the microcontroller and to debug the microcontroller. If you do not want a USB port or RS-232 DBGU port on the board, the JTAG connection is the only other way to load a program to the microcontroller while it is on the board. It is also possible to load a program to the microcontroller Flash memory with a chip programmer before soldering it to the board.

USB port: A USB device port is shown connected to the microcontroller. This port will only work if the microcontroller is operating from an 18.432MHz crystal and the master clock is set to 48MHz in software. Two additional microcontroller pins are used with the USB. PA16 is used to connect a 1k5 pull-up resistor to the USB D+ line. When Q1 is switched on by the microcontroller, it pulls up the D+ line which notifies the USB host that a USB device is connected to it. An additional port pin PA31 is used to sense when the USB port is plugged into a host. This is not necessary for the on-chip SAM-BA boot loader to work.

DBGU RS-232 port: Port pins PA10 and PA9 connect to the internal DBGU UART. A RS-232 level converter chip (U2) that operates from 3.3V is used to interface the UART to an external RS-232 port.

Four I/O pins are used in the schematic. If you wanted the full 32 port pins available for your own use and still want to load programs into the microcontroller while it is soldered to the board, it can be done in one of two ways. Firstly via a JTAG ICE connected to the JTAG port or secondly via the USB port with a permanent 1k5 pull-up resistor connected to the D+ USB line (microcontroller DDP pin).

When using the JTAG port, any crystal value between 3 and 20MHz can be used with the microcontroller. An 18.432MHz crystal is always required when using the USB port. The SAM-BA boot loader will still work without using port pin PA16 to switch the 1k5 resistor onto the D+ USB line. I have tried this and found that it does work, but may confuse the PC USB device detection when it is plugged into the PC and a microcontroller program is not run immediately that starts the USB negotiation protocol. The PC knows that a USB device is connected because it detects the 1k5 resistor, but cannot communicate with it. If the SAM-BA boot program is loaded in the microcontroller Flash, it starts running immediately when plugged into the USB port if it is powered by the USB 5V. The USB negotiation with the PC then takes place within the available time for USB transactions to take place.

Note that the circuit of **figure 1.8** does not require any reset circuitry on the NRST reset pin of the microcontroller. Reset circuitry is built into the microcontroller including power supply brown-out detection. In fact the reset pin must be enabled in software in order for an external reset switch to be able to reset it.

1.8 References

This section provides various documents and website references that provide additional information on the topics covered in this chapter.

Software toolchain and debugging

1. **www.yagarto.de** – contains the tutorial on installing and using the YAGARTO toolchain.

2. **Tutorial by James P. Lynch** – this is probably the most important document to read as it contains information on using the YAGARTO toolchain including using the SAM-ICE for debugging. The YAGARTO installation instructions in this tutorial are a bit out of date as YAGARTO has changed since the tutorial was written, but the rest of the tutorial still applies. There is a link to this tutorial on the YAGARTO front page. Or here:
   www.atmel.com/dyn/resources/prod_documents/atmel_tutorial_source.zip
   The zipped file contains the tutorial in pdf format as well as example programs.

Atmel AT91SAM7S-EK evaluation kit

1. **Atmel evaluation kit page** at **www.atmel.com**:
   www.atmel.com/dyn/products/tools_card.asp?tool_id=3784

2. **AT91SAM7S-EK Evaluation Board User Guide (doc6112.pdf)** – can be downloaded from the above page.

3. **AT91SAM7S-EK Software Package for IAR 5.2, Keil and GNU (at91sam7s-ek.zip)** – software example programs for use with the evaluation kit. Can be downloaded from the above page.

AT91SAM7S datasheets

1. The **datasheets for the AT91SAM7S** microcontrollers can be downloaded at:
   www.atmel.com/dyn/products/product_card.asp?part_id=3524

2. **AT91SAM7S Series Preliminary Summary (6175s.pdf)** - contains introductory information on the AT91SAM7S family.

3. **AT91SAM7S Series Preliminary (doc6175.pdf)** – is the full datasheet for the AT91SAM7S family.

Hardware design with the AT91SAM7S

1. **AT91SAM7S-EK schematic** – the schematic diagram for the AT91SAM7S-EK is available at the back of the **AT91SAM7S-EK Evaluation Board User Guide (doc6112.pdf)**

2. **Application notes** – available from the Atmel website:
   www.atmel.com/dyn/products/product_card.asp?part_id=3524

3. Also look under "**Other Related Application Notes**" on the above page.

Forum and Community Website for AT91SAM Microcontrollers

1. **www.at91.com**

Commercial Software Tools

1. **www.keil.com** – commercial C/C++ compilers for ARM and other architectures.
2. **www.iar.com** – commercial C/C++ compilers for ARM and other architectures.
3. **www.rowley.co.uk** – C/C++ GNU compilers for ARM and other architectures. Includes their own IDE and C runtime library. Have commercial, educational and personal licenses.
4. **www.codesourcery.com** – C/C++ GNU compilers for ARM and other architectures. Sourcery G++ Lite Edition available for free. Sourcery G++ includes GNU C/C++ compilers, Eclipse IDE, runtime library and support for tools. Offer personal, standard and professional editions.

5. Other commercial software tools are available for ARM microcontrollers, but the list above covers the compilers for which Atmel provide source code example programs in their software package for evaluation boards. Examples are provided for IAR, Keil and GNU toolchains.

This chapter contains many details that may not be needed immediately, such as information on designing a new microcontroller circuit. The simplest way to start using AT91SAM7S microcontrollers or indeed any microcontroller if you have an evaluation board is:

- Read the evaluation board documentation, power up the evaluation board to see that it is working. Evaluation boards usually have a default program loaded on them when purchased. The evaluation board documentation should tell you what it is and how to re-load it if it has been erased.

- Find out how to load already-compiled binary programs to the board and see that they work. On the AT91SAM7S-EK this involves installing the SAM-BA program on the PC and then the on-chip SAM-BA program on the board using the TST jumper, then connecting via USB, DBGU serial port or JTAG port (if you have a SAM-ICE) using SAM-BA.

- After you have successfully loaded example programs to the evaluation board, install the software toolchain. Test the toolchain by opening a project, compiling it and loading the binary output file to the evaluation board.

- Try modifying a program, compiling and loading it to the board.

- Try creating your own program from scratch, compile it and download it to the board.

- Once you have the toolchain running, can create and compile your own programs, and load the programs to the evaluation board, you can try to interface a hardware device such as a simple LED to the board and move to more complex devices from there.

Now that you have a hardware and software platform to work from, we can start interfacing to the microcontroller starting from the next chapter.

2. Digital Outputs

Digital outputs refer to the general purpose output pins of the microcontroller and are also known as ***discrete outputs***. Most microcontrollers have input/output (I/O) pins that can be configured as either inputs or outputs or can be connected to one of the internal peripheral devices in the microcontroller.

This chapter will show how various devices can be interfaced to the microcontroller I/O pins that are configured as outputs. At the end of the chapter, software examples are given on how to program the I/O pins as outputs and switch them on and off to control the interfaced devices.

This chapter will show how to interface the following devices:

- LEDs
- Bipolar Junction Transistors
- MOSFETs
- Optocouplers
- Power Transistors (BJT and MOSFET)
- Relays and Solenoids
- D.C. Motors (on and off control, not speed control)

Other devices such as LCD and 7-segment displays are also interfaced to the digital outputs of the microcontroller, but are covered in a separate chapter.

Microcontroller output pin current delivery capabilities:
Before connecting any device to a microcontroller pin that is set up as an output, it is important to know how much current the microcontroller pins can deliver (source or sink). This information is available in the electrical characteristics section of the microcontroller data sheet. **Table 2.1** shows the current delivery capabilities of four ARM microcontrollers.

	Atmel ARM Microcontrollers		NXP ARM Microcontrollers	
	AT91SAM7S256	**AT91SAM7A3**	**LPC2148**	**LPC2138**
Current per pin	PA0 to PA3 16mA PA17 to PA20 2mA All other I/O pins 8mA	2mA	4mA	4mA
Total current all pins	150mA Absolute Maximum	200mA Absolute Maximum	Not Specified	Not Specified

Table 2.1 – Microcontroller Pin Output Currents

Something to look out for when using the pins on a microcontroller as outputs is not to exceed the maximum current that can be drawn from all of the pins together. Take the example of the **AT91SAM7S256** in **Table 2.1**. If we set up all 32 I/O pins as outputs and draw the maximum current from each, we will draw 264mA (4 ×16mA + 4 × 2mA + 24 × 8mA) which is far greater than the 150mA absolute maximum that can be drawn from all pins. So be sure to calculate the total current drawn from each pin and make sure that it fits well below the absolute maximum rating that is specified.

The **AT91SAM7A3** has 62 I/O pins, so if they were all set up as outputs and the maximum current per pin was drawn, we would draw 124mA (62 × 2mA) which fits well within the 200mA absolute maximum current.

The **LPC2148** and **LPC2138** current rating per pin is actually specified as a minimum of 4mA each. We have to assume that we can draw 4mA from each I/O pin simultaneously.

All microcontrollers in **Table 2.1** can *source* or *sink* the rated current per pin. Older microcontrollers can normally sink more current than they can source, so always check this in the datasheet.

Current Sourcing – Current flows from the device (microcontroller pin in this case) into a load attached to it.

Current Sinking – Current flows into the device (microcontroller) from the load that is attached to it.
See the LED interface examples below to better understand current sinking and sourcing.

Microcontroller I/O pin state at reset:
It is important to know what state the I/O pins of the microcontroller are in at reset (e.g. when power is first applied to the microcontroller). The AT91SAM7S microcontroller's I/O pins default to inputs with pull-up resistors enabled. (Each I/O pin has an individually controlled pull-up resistor that can be enabled or disabled in software). If the pull-up resistor is enabled, it may switch the device that is attached to it on or partly on which may or may not be a problem depending on the design requirements of the embedded system.

The software program that starts running when the microcontroller is switched on can disable the pull-up resistor at the beginning of the program, however this will only be after the microcontroller has booted up and run some initialisation code. For an embedded system that is safety critical, for example if the microcontroller output starts a motor that could potentially cause some damage, interface circuitry must be built to keep the motor off in the event that the microcontroller remains in a reset state for some reason. In order to design a safe interface circuit, it is essential to know what the state of the microcontroller pin will be at boot-up (reset) – also consider what will happen if a reset button is made available to the user and can be held in by the user.

Another important consideration at reset is to make sure that the circuitry connected to the microcontroller pins does not pull the pins to an undesired state. For example, to use the on-chip boot loader of the AT91SAM7S, pins PA0 to PA2 must be pulled high when the TST pin is pulled high in order to load the on-chip boot program to Flash memory. If any circuitry is attached to one of these pins that pulls any one of them low, the boot program will not load. If the boot program is not required, however, then this is not a problem. On a microcontroller, such as the NXP LPC2148, there are many more pins that must be pulled to a certain logic level at reset to enable or disable various devices, so be sure to read the datasheet carefully.

2.1 Interfacing LEDs

The Light Emitting Diode (LED) is probably the simplest device that can be interfaced to a microcontroller. When starting with a new microcontroller or when testing a newly built microcontroller board, one of the first tests done to see if the board is working is to flash an LED. If you can flash an LED on and off that is attached to a microcontroller I/O pin, then you have got the software toolchain up and running, you can load the program into the microcontroller and you know that the microcontroller is working, or at least the microcontroller core and an I/O pin is working.

Uses for LEDs

- As an indicator when debugging.
- To show communications activity on, for example, a serial port to show that data is being sent and/or received.
- To display status information of the embedded system.
- Anywhere that visual information is to be displayed to a user.
- To display numbers such as a count value or time when the LEDs are arranged in a seven segment display.
- Infrared LEDs for communications such as IR remote controls.

2.1.1 LED Interfacing Quick Reference

Figure 2.1 shows the two ways of interfacing an LED directly to a microcontroller pin. The dotted line in the figure shows the direction of current flow for each configuration when the LED is switched on. V_{CC} in the figure will be the I/O voltage of the microcontroller in use e.g. V_{CC} = 3.3V for a microcontroller with 3.3V I/O such as the AT91SAM7S. **Figure 2.2** shows examples of a through hole and surface mount LED and the polarity that must be observed when connecting an LED.

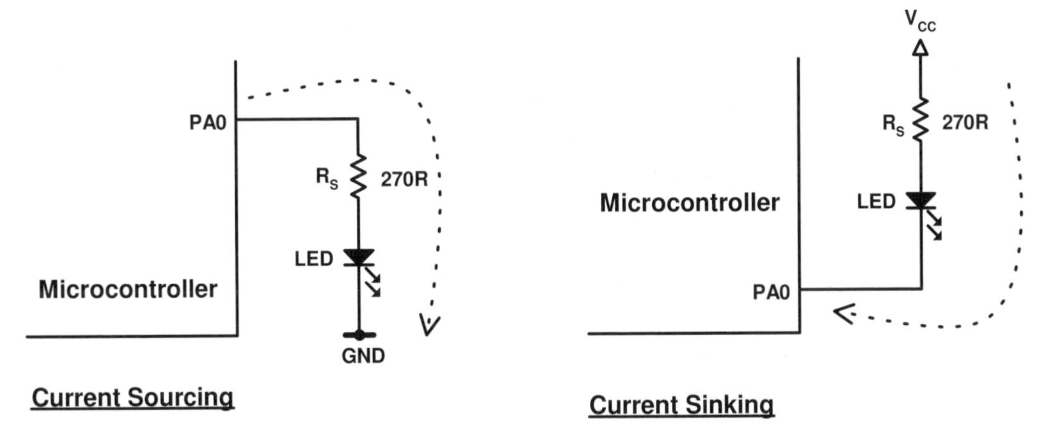

Figure 2.1: Interfacing an LED to the Microcontroller I/O Pin

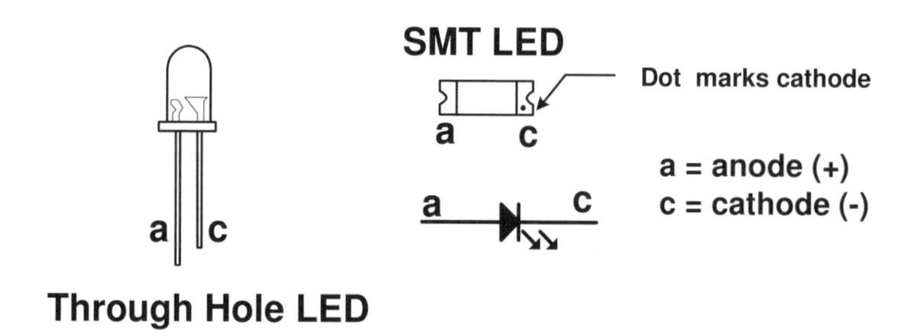

Figure 2.2: LED Polarity

Some common LED series resistor values for 3.3V and 5V systems are shown in **table 2.2** for LEDs with a 1.7V forward voltage drop. **N.B.:** First check that the microcontroller pin can deliver the desired current.

3.3V System		5V System	
Resistor (R_S)	**LED Current**	**Resistor (R_S)**	**LED Current**
180Ω	8.9mA	270Ω	12.2mA
270Ω	5.9mA	330Ω	10mA
680Ω	2.4mA	470Ω	7mA
1k	1.6mA	560Ω	5.9mA

Table 2.2: Some Commonly Used LED Series Resistor Values

Choosing between a current sourcing and current sinking LED configuration:
In a ***current sourcing*** configuration, logic 0 on the microcontroller pin switches the LED off and logic 1 switches the LED on. See the software section at the end of this chapter for information on how to write software to control the LED.
On AT91SAM7S microcontrollers, the internal pull-up resistors on the pins that are enabled at reset will supply enough current that will cause the LED to glow dimly at reset in a current sourcing configuration.

When the LED is wired in ***current sinking*** configuration, a logic level of 0 on the microcontroller pin switches the LED on and a logic level of 1 switches the LED off.

On AT91SAM7S microcontrollers, the internal pull-up resistors on the pins that are enabled at reset will keep the LED off in current sinking configuration at power-up.

What if the microcontroller pin cannot deliver enough current to drive the LED?
You will need to drive the LED with a transistor. See the next section "**Interfacing Transistors**".

2.1.2 LED Interfacing Explanation and Calculations

An LED must be connected to the microcontroller pin through a series resistor in order to limit the current through the LED. If a series resistor is not used, either the LED would burn out or the microcontroller circuitry that drives the pin could blow.

When connecting up LEDs in a circuit, the correct polarity must be observed. The anode of the LED must be connected to the positive (+) side of the circuit and the cathode must be connected to the negative (-) or 0V (GND) side of the circuit as shown in **figure 2.1** and **figure 2.2**.

The current limiting resistor:

1) Controls the brightness of the LED.
2) Prevents more than the maximum current that the microcontroller pin can source/sink from being reached.
3) Prevents more than the maximum forward current that the LED can handle from being reached.

To calculate the value of the series resistor used with the LED:

1. Choose the LED forward current (I_F)

- Choose between **2mA to 10mA** depending on the brightness required.

- The chosen current is limited by the amount of current that the microcontroller pin can source or sink and the maximum forward current of the LED. Check the microcontroller and LED datasheets for these values.

The value of the series resistor is calculated by first deciding how much current will be allowed to flow through the LED. Most general purpose LEDs will glow at only 1.4mA although they may not be too bright. A blue LED that I tested glowed fairly brightly at only 0.7mA. And a clear red hyper-bright LED also glowed fairly brightly at 1.4mA.

Most general purpose LEDs will operate safely with between **1.4mA to 20mA** of forward current. Always be sure to check the datasheet to see that the LED that you are using operates in this range. The higher the forward current, the brighter the LED will glow. Most designs that I have seen that drive an LED directly from a microcontroller pin use between **2mA and 10mA** which I think is a good range to work with. Remember to always check if your microcontroller pin can deliver the chosen current.

2. Get the LED forward voltage (V_F)

- Get this value from the LED datasheet.

- If the datasheet is not available, use **1.7V** although this value may vary widely between different coloured LEDs.

The LED forward voltage is the voltage that will be dropped across the LED when it is switched on (forward biased). This information is available from the LED manufacturer's datasheet. Different coloured LEDs have different forward voltage drops, so it is important to get this value from the manufacturer, although a value of **1.7V** is probably the best rule of thumb if you can't get the datasheet (I have seen various text books using values between 1.6V and 1.8V). This value can be used for general purpose LEDs but not for high power LEDs. Another way to get the voltage drop across the LED is to set it up with a series resistor on a breadboard and measure the voltage drop across it. Some LEDs will have a 2.2V or greater forward voltage drop. Using the 1.7V rule of thumb value for an unknown LED is safer than using a bigger value because the calculated resistor value will allow less current to flow, preventing the possibility of drawing too much current from a microcontroller pin.

3. Calculate the voltage that will be dropped across the series resistor (R_S)

Subtract the LED forward voltage from the microcontroller I/O pin voltage to get this value. The microcontroller I/O pin voltage will also be the same as V_{CC} in **figure 2.1**. E.g. in a 3.3V ARM microcontroller such as the AT91SAM7S this will be 3.3V (V_{CC} and V_{IO}).

$$\mathbf{V_{RS} = V_{CC} - V_F}$$

OR

$$\mathbf{V_{RS} = V_{IO} - V_F}$$

Where:
$\mathbf{V_{RS}}$ = Voltage dropped across the series resistor R_S
$\mathbf{V_{CC}}$ = $\mathbf{V_{IO}}$ = Microcontroller supply voltage = Microcontroller I/O voltage
$\mathbf{V_F}$ = Forward voltage drop across the LED

E.g. in a 3.3V system with a LED V_F of 1.7V:

$$\begin{aligned} V_{RS} &= V_{CC} - V_F \\ &= 3.3V - 1.7V \\ &= 1.6V \end{aligned}$$

4. Calculate the value of the series resistor using Ohm's Law:
$\mathbf{R_S = V_{RS} \div I_F}$

E.g. with a chosen forward current of 8mA:

$$\begin{aligned} R_S &= V_{RS} \div I_F \\ &= 1.6V \div 8mA \\ &= 200\Omega \end{aligned}$$

5. Calculate the power rating of the resistor:
It is important to check that the resistor that you are going to be using has a big enough power rating to handle the power that will be dissipated in it when the LED is on.

$$\begin{aligned} P &= I^2R \\ &= (8mA)^2 \times 200\Omega \\ &= 12.8mW \end{aligned}$$

Or use:

$$\begin{aligned} P &= V \times I \\ &= 1.6V \times 8mA \\ &= 1.6 \times 0.008 \\ &= 0.0128 \\ &= 12.8mW \end{aligned}$$

The most common resistors around are ¼W resistors (**¼W = 250mW**) which is more than enough for this example. Surface mount resistors with a 1206 footprint are normally rated at ¼W, 0805 footprint resistors are normally rated at ⅛W (**⅛W = 125mW**).

If you do not have the calculated resistor:
If you do not have a 200Ω resistor, you will need to choose the closest value that you do have in stock and see if the current flow through the LED is acceptable. For example, say that the closest value that you have to 200Ω is 180Ω. Now calculate how much current will flow through the LED using this value of resistor and see if it is acceptable.

```
I = V ÷ R
  = 1.6V ÷ 180Ω
  = 0.00889A
  = 8.89mA
```

You must now decide if 8.89mA is acceptable or if you must rather choose a bigger value resistor or go out and buy some 200Ω resistors.

2.1.3 Some Measured LED Voltages

A range of different LEDs were attached to one of the high current drive pins of an AT91SAM7S microcontroller as shown in **figure 2.3** and the LED switched on using software (PA0 to PA3 are high current drive pins). The test results using a 330Ω and 150Ω series resistor in current sinking configuration are shown in **table 2.3**.

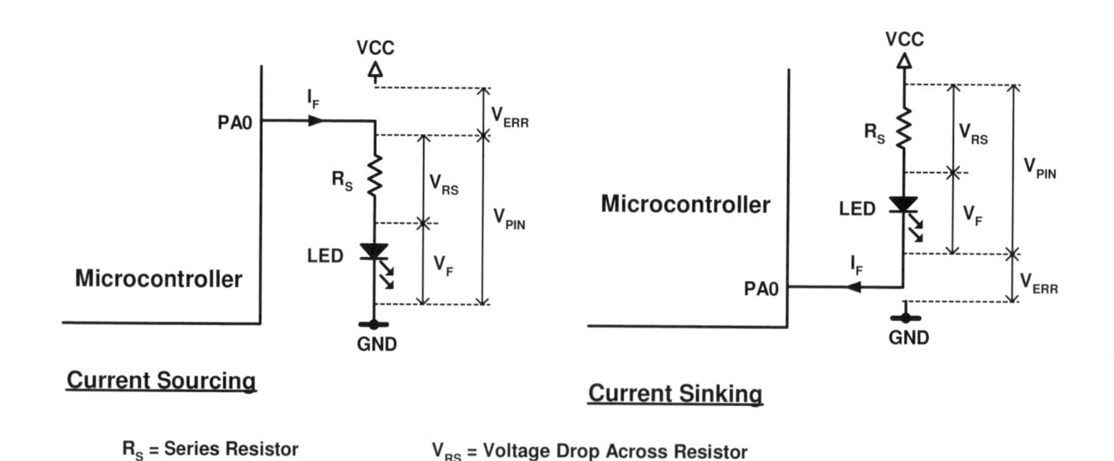

Figure 2.3: Measurements Taken on an AT91SAM7S Microcontroller Interfaced to an LED

	$R_S = 330\Omega$				
LED	**V_F**	**V_{RS}**	**V_{ERR}**	**V_{PIN}**	**I_F**
Calculated	1.7V	1.6V	0V	3.3V	4.85mA
3mm RED	1.91V	1.35V	0.04V	3.26V	4.1mA
3mm GREEN	1.98V	1.29V	0.04V	3.26V	3.9mA
3mm YELLOW	1.95V	1.31V	0.04V	3.26V	4mA
3mm RED BRIGHT	1.95V	1.31V	0.04V	3.26V	4mA
5mm RED	1.71V	1.55V	0.04V	3.26V	4.7mA
5mm GREEN	1.98V	1.28V	0.04V	3.26V	3.9mA
5mm AMBER	1.47V	1.47V	0.04V	3.26V	4.5mA
5mm BLUE	2.68V	0.6V	0.02V	3.28V	1.8mA
5mm YELLOW	1.99V	1.28V	0.04V	3.26V	3.9mA
5mm RED BRIGHT	1.69V	1.57V	0.04V	3.26V	4.8mA
	$R_S = 150\Omega$				
LED	**V_F**	**V_{RS}**	**V_{ERR}**	**V_{PIN}**	**I_F**
Calculated	1.7V	1.6V	0V	3.3V	10.67mA
3mm RED	1.99V	1.24V	0.07V	3.23V	8.3mA
3mm GREEN	2.06V	1.17V	0.07V	3.23V	7.8mA
3mm YELLOW	2.06V	1.17V	0.07V	3.23V	7.7mA
3mm RED BRIGHT	2.03V	1.17V	0.07V	3.23V	7.9mA
5mm RED	1.75V	1.47V	0.08V	3.22V	9.7mA
5mm GREEN	2.07V	1.16V	0.08V	3.22V	7.7mA
5mm AMBER	1.88V	1.34V	0.08V	3.22V	8.9mA
5mm BLUE	2.75V	0.52V	0.03V	3.27V	3.5mA
5mm YELLOW	2.12V	1.11V	0.07V	3.23V	7.4mA
5mm RED BRIGHT	1.72V	1.49V	0.09V	3.21V	9.9mA

Table 2.3: Test Results of Various LEDs Interfaced to an AT91SAM7S Microcontroller – current sinking configuration

V_{ERR} is the difference between the supply rail (0V or 3.3V) and the voltage that the pin switches to when trying to switch to the supply voltage. When switching the LED on in current sourcing mode, the pin tries to switch to V_{CC} (3.3V on the AT91SAM7S) but falls a bit short because the switching circuitry in the microcontroller is not a perfect conductor. In a current sinking configuration, the pin tries to switch to 0V to switch the LED on, but cannot reach this exact value.

LED interfacing notes:

- The same coloured LEDs from different manufacturers may have different forward voltage drops.
- Different coloured LEDs have different forward voltage drops.
- The microcontroller pin cannot switch exactly to 0V to switch the LED on in a current sinking configuration.
- When the value of the series resistor is decreased i.e. the forward current of the LED is increased, V_F increases – this can also be seen in the LED datasheet.
- When the LED forward current is increased, the microcontroller pin voltage that switches the LED on is slightly further from the rail voltage that it is trying to switch to. This value is still very small, i.e. less than 0.1V in the table, so is insignificant.
- The calculated forward LED current is conservative and calculates a higher current than is measured. This is due to the forward LED voltage being estimated as 1.7V.
- The LED forward voltage in the datasheet is usually specified as a typical value and a maximum value with the maximum value being significantly higher than the typical value. Strictly speaking current calculations should be made for the lowest LED forward voltage and highest LED forward voltage specified in the datasheet to see what the range of possible currents could be drawn when using a specific resistor value.

Table 2.4 compares two LEDs in current sourcing and current sinking mode on the AT91SAM7S (V_{CC} = 3.3V). As can be seen, a slightly higher V_{ERR} in current sourcing mode results in a slightly lower LED forward current. V_{ERR} is bigger in current sourcing mode because the internal switching circuitry of the AT91SAM7S can switch the voltage of the pin closer to the 0V rail when switching to logic low than to the 3.3V rail when switching to logic high.

Most engineers will design using a current sinking configuration as older logic ICs could sink more current than they could source, even though the AT91SAM7S can source and sink equal currents. When the AT91SAM7S boots up or is reset, the I/O pins are configured as inputs with pull-up resistors enabled. In current sourcing configuration, the LED will glow dimly if a program does not run that switches the LED off. In current sinking configuration, the LED will stay off at reset. See the software section at the end of this chapter for information on enabling and disabling internal pull-up resistors on the AT91SAM7S.
Any discrepancies between the calculated currents in **tables 2.3** and **2.4** and the measured currents are due to the tolerance (accuracy) of the resistor and/or rounding off error of the multimeter used to measure the voltages and currents in the circuit. A resistor with a tolerance of 5% was used. E.g. in **table 2.3**, when the 150Ω resistor has a voltage of 1.47V across it, the calculated current is 9.8mA, but the measured current is 9.7mA.

Current Sinking, R_S = 150Ω					
LED	**V_F**	**V_{RS}**	**V_{ERR}**	**V_{PIN}**	**I_F**
3mm RED	1.99V	1.24V	0.07V	3.23V	8.3mA
5mm RED	1.75V	1.47V	0.08V	3.22V	9.7mA
Current Sourcing, R_S = 150Ω					
LED	**V_F**	**V_{RS}**	**V_{ERR}**	**V_{PIN}**	**I_F**
3mm RED	1.98V	1.22V	0.09V	3.21V	8.1mA
5mm RED	1.75V	1.44V	0.11V	3.19V	9.5mA

Table 2.4: Test Results of the Same Two LEDs in a Current Sinking and Current Sourcing Configuration

2.2 Interfacing Transistors

A transistor can be used as a switch to switch larger currents than a microcontroller pin can. This enables the microcontroller to switch on devices such as relays, solenoids, motors, light bulbs and other ***direct current (d.c.)*** operated devices.

This section shows the use of low power general purpose transistors to drive LEDs. **Section 2.4** in this chapter shows how to drive other devices such as relays. Bipolar Junction Transistors (BJT) and Field Effect Transistors (FET) are discussed.

Uses for transistors

To drive loads that require larger currents than a microcontroller pin can deliver such as:

- Relays
- Solenoids
- LEDs
- Motors

2.2.1 Bipolar Junction Transistors (BJT)

2.2.1.1 Transistor Interfacing Quick Reference

Figure 2.4 shows how to use a NPN BJT to drive an LED at higher currents than a microcontroller pin can on its own. The microcontroller switches the transistor on and the transistor switches the LED on. Note that the transistors in the two circuits in the figure are switching different voltages. The first is operating from the microcontrollers 3.3V supply, the second is operating from an external 12V supply.

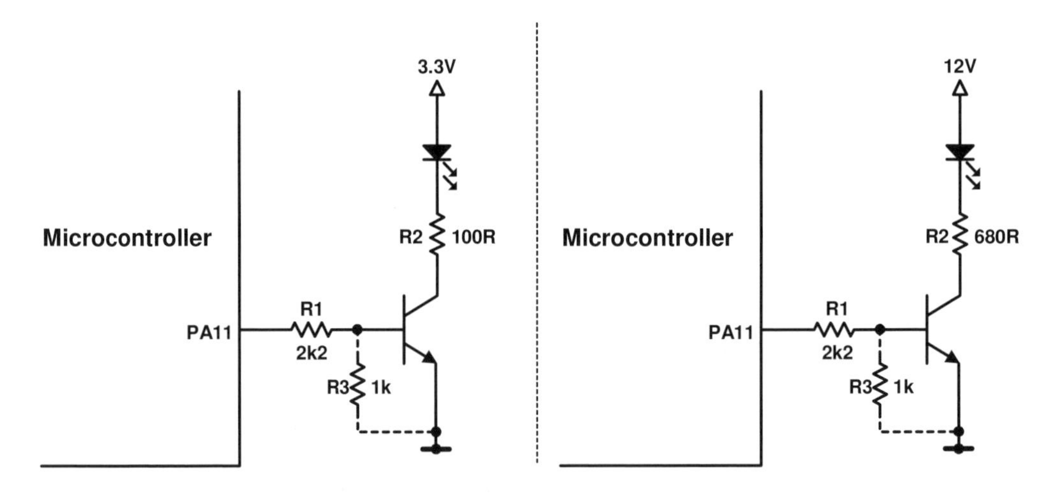

Figure 2.4: NPN Transistor Used as a Switch by a Microcontroller

The left circuit in **figure 2.4** drives the LED with 14mA of current from a microcontroller pin that could normally only drive an LED with 8mA maximum. To switch the transistor on, the microcontroller pin delivers just over 1mA of current. The right circuit drives the LED with nearly 15mA from a 12V power supply.

Figure 2.5 shows some commonly used NPN transistors and their packages. These general purpose transistors are well suited to being used as switches.

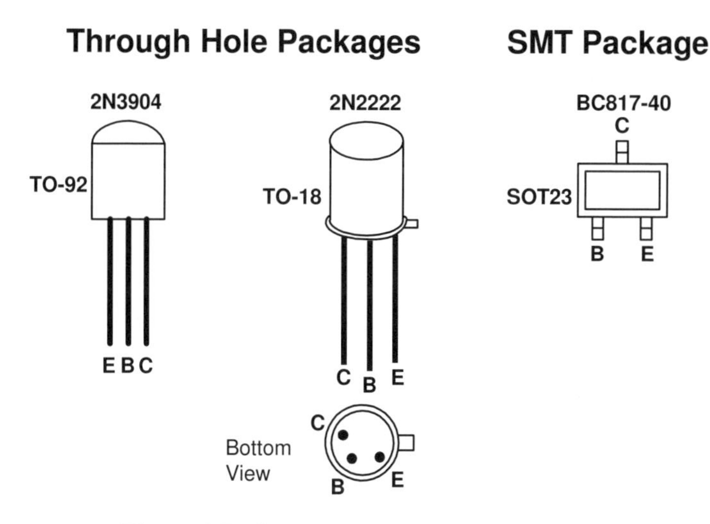

Figure 2.5: General Purpose NPN Transistors

Using any of the transistors in **figure 2.5** with the circuits of **figure 2.4** allows driving a load with up to **60mA** of current easily and safely from a 3.3V microcontroller. These same transistors could drive even more current by using a different base resistor as they are rated for driving higher currents.

Note that because AT91SAM7S microcontrollers have an internal pull-up resistor enabled at reset, the transistor in this circuit will switch on when the circuit is powered on. The optional pull-down resistor, R3 in **figure 2.4**, will keep the transistor off at power-up. This resistor has to be fairly strong (1k) to keep the transistor off. A 2k2 resistor was tested and it was not strong enough to keep the transistor fully off, the LED glowed dimly with the 2k2 pull-down resistor.

2.2.1.2 Transistor Interfacing Explanation and Calculations

How a transistor operates as a switch:

In order to switch a NPN transistor on, a voltage must be applied to the base of the transistor that will cause enough current to flow through the base-emitter junction of the transistor. When this base-emitter current is flowing, the transistor will switch on i.e. it will conduct a current from the collector to the emitter.

In the left half of **figure 2.6**, the microcontroller is applying a logic '0' to the base of the transistor (①). The base of the transistor is now at the same voltage potential as the emitter. No current will flow from the base to the emitter (②) which will keep the transistor off. No current will flow from the collector to the emitter (③) which keeps the LED off.

In the right half of **figure 2.6**, the microcontroller applies a logic '1' to the base of the transistor (①). This causes a current to flow from the base of the transistor to the emitter (②), which switches the transistor on, causing a current to flow from the collector of the transistor to the emitter of the transistor (③). This in turn switches the LED on.

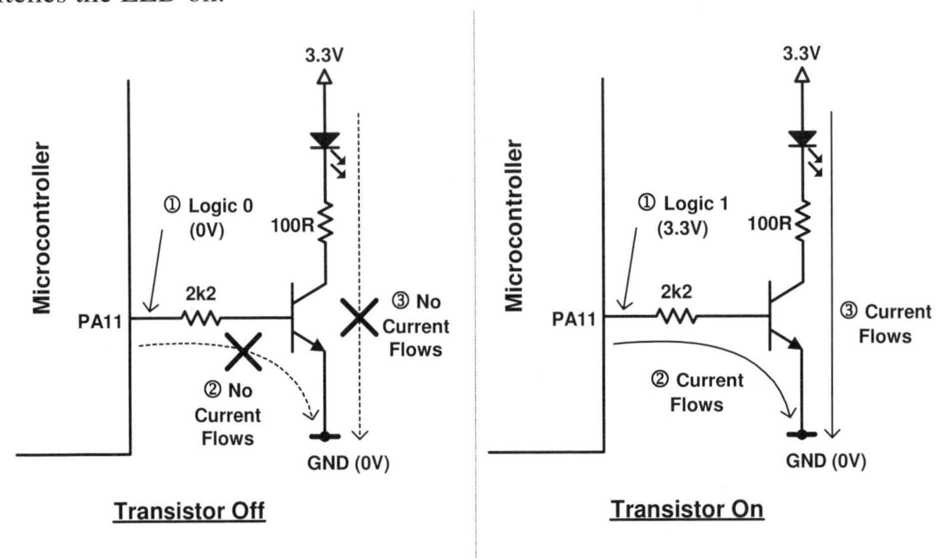

Figure 2.6: How a NPN Transistor Operates as a Switch

Transistor calculations:

Refer to **figure 2.7** when reading this section.

When switched on the base-emitter junction of the transistor is like a forward biased diode and will therefore have 0.7V (V_{BE}) dropped across it (some text books use 0.6V for this value). The remainder of the voltage will be dropped across the base resistor of the circuit (R_B). In a 3.3V system, this leaves 2.6V across the base resistor (3.3V – 0.7V).
We must now decide how much current to allow to flow through the base of the transistor and then use Ohm's Law to calculate the base resistor.

The amount of current allowed to flow through the base of the transistor is limited by the amount of current that the microcontroller pin can deliver as well as the amount of base current that the transistor can handle (found in the transistor datasheet).

The gain of the transistor is used to calculate how much base current (I_B) will switch the transistor fully on (this is known as 'saturation'). The gain of the transistor being used can be found in the transistor datasheet and is represented by the symbol β (beta) or h_{FE}.

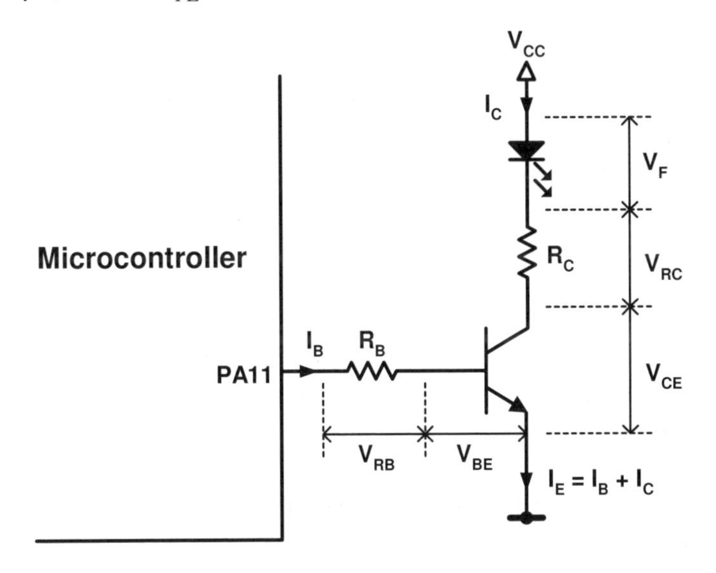

R_B = Base Resistor

R_C = Collector Resistor

I_B = Base Current

I_C = Collector Current

I_E = Emitter Current

V_{RB} = Voltage across Base Resistor

V_{RC} = Voltage across Collector Resistor

V_F = LED Forward Voltage

V_{CE} = Transistor Collector Emitter Voltage

V_{BE} = Transistor Base Emitter Voltage

Figure 2.7: NPN Transistor Parameters

Base current and resistor calculation example:

First calculate how much collector current (I_C) will be needed. Let's say that we want to drive our LED with 16mA, we will therefore need a collector current of 16mA. Now look in the transistor datasheet to see what the minimum gain of the transistor is, for this example say 100.

We now have:
$I_C = 16mA$
$h_{FE} = 100$

The formula to calculate the base current is:

$$\begin{aligned} I_B &= I_C \div h_{FE} \\ &= 16mA \div 100 \\ &= 0.016 \div 100 \\ &= 0.00016 \\ &= 160\mu A \end{aligned}$$

160µA is the minimum current required to switch the transistor fully on, also known as saturating the transistor. It is good practice to increase the calculated base current of the transistor by about **30%** to ensure that it stays in saturation state reliably. This compensates for component tolerances and slight variation in operating parameters such as voltage and temperature.

In an industrial environment (electrically noisy environment) a base current of at least **1mA** should be used to prevent external interference from switching the transistor off, even if momentarily. Noise could intermittently switch the transistor off causing spikes on the load that it is driving.

Good design practice for industrial environments:
Use a base current of at least 1mA.
OR
If the calculated base current is 1mA or more then use this current plus at least 30%.

In our example, I_B will be therefore be:
$160\mu A \times 1.3 = 208\ \mu A$
OR
1mA
For industrial environments.

In a 3.3V system, there will be 2.6V dropped across the base resistor (3.3V – 0.7V), so using Ohm's Law:

```
R = V ÷ I
  = 2.6V ÷ 208µA
  = 2.6 ÷ 0.000208
  = 12500Ω (12k5)
```

Choose a standard 12k resistor or even 10k for this example.

Or using 1mA for the base current:

```
R = V ÷ I
  = 2.6V ÷ 1mA
  = 2.6 ÷ 0.001
  = 2600Ω (2k6)
```

A 2k2 resistor could be used and will allow 1.18mA to flow through the base of the transistor:

```
I = V ÷ R
  = 2.6V ÷ 2k2
  = 2.6 ÷ 2200
  = 0.00118 (1.18mA)
```

To calculate the maximum collector current that the load could draw:

$$I_B = I_C \div h_{FE}$$

therefore

$$\begin{aligned} I_C &= I_B \times h_{FE} \\ &= 1.18\text{mA} \times 100 \\ &= 0.00118 \times 100 \\ &= 0.118 \\ &= 118\text{mA} \end{aligned}$$

Note that this is the maximum value and it would be better to use a load that draws much less current than this maximum, say only 80mA. This same calculation was used to calculate the maximum collector current of the example circuits in **figure 2.4**, but using a minimum gain of 70 instead of 100.

To calculate the LED series resistor (collector resistor, R_C) we subtract 1.7V and 0.2V from the supply voltage (3.3V in this example) to find out how much voltage is dropped across the resistor. The 0.2V is the voltage that will appear across the collector-emitter junction of the transistor when it is fully on (V_{CE}) and the 1.7V is the LED forward voltage (V_F).

If we want drive the LED at 16mA, the calculation will be:

$$V_{RC} = V_{CC} - V_F - V_{CE} = 3.3V - 1.7V - 0.2V = 1.4V$$

$$\begin{aligned} R &= V \div I \\ &= 1.4V \div 16mA \\ &= 1.4 \div 0.016 \\ &= 87.5\Omega \end{aligned}$$

Again this resistor value will need to be rounded to the nearest standard value.

A power calculation must be done to see what power rating our resistor must be:

$$\begin{aligned} P &= VI \\ &= 1.4V \times 16mA \\ &= 1.4 \times 0.016 \\ &= 0.0224 \ (22.4mW) \end{aligned}$$

So a ¼W or ⅛W resistor will be more that enough.

Interfacing to a load with a different voltage to the microcontroller:

It is possible to drive a load that operates at a different voltage to the microcontroller I/O pins. **Figure 2.8** shows an LED being driven by an external 12V d.c. power supply.

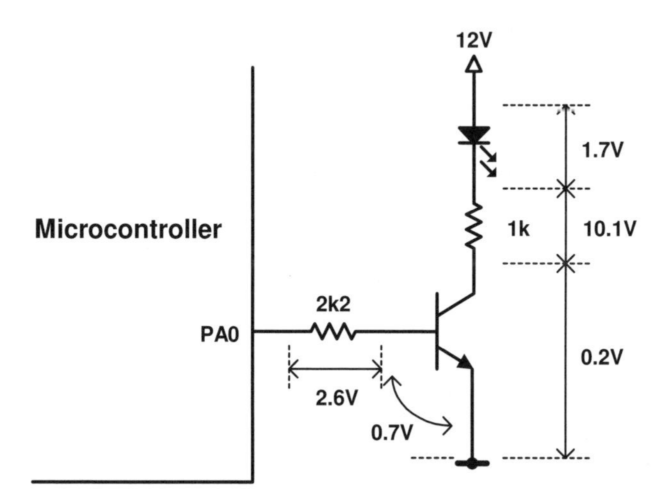

Figure 2.8: A Transistor Switching a Load that Operates from an External 12V Supply

Transistors can only be used to switch a d.c. load. The voltage that can be switched depends on the transistor selected and this voltage can be found in the transistor manufacturer's datasheet. When switching very high voltages, it is better to use an optocoupler or relay or both (see the sections that follow in this chapter).

When using an NPN transistor as in **figure 2.8**, the microcontroller only has to pass a big enough current through the base emitter junction to switch the transistor on. The figure shows a 3.3V system driving an LED that operates from an external 12V supply.

The LED series resistor or collector resistor shown in the 12V circuit of **figure 2.4** and **figure 2.8** must use a ¼W (250mW) or higher rated resistor as the power that is dissipated in this resistor will be 150mW and 102mW which is too high for a ⅛W (125mW) resistor (remember to always design well below the components rating for reliability).

Keeping the transistor off at power-up or reset:

As AT91SAM7S microcontrollers enable the internal pull-up resistors on each pin at reset, an NPN transistor interfaced to one of the microcontroller pins will switch on when the microcontroller system is powered up or reset. Software can be written to immediately switch the transistor off at start-up, but this may not be convenient if, for example, a motor is being driven by the transistor. If the microcontroller is then held in reset for some reason, the motor would run and possibly cause damage.

Another solution is to use a **pull-down resistor** on the base of the transistor to pull the base to 0V and keep the transistor off at power-up as shown in **figure 2.4** – resistor R3 keeps the transistor off.

Another solution is to use a **PNP transistor** as shown in **figure 2.9**, however this solution must only be used when switching a load that uses the microcontrollers I/O voltage. If, for example, the load was being operated from 5V (V_{CC} = 5V in the figure) and the microcontroller I/O voltage is 3.3V two things will happen. Firstly while the internal pull-up resistor on an AT91SAM7S is enabled, most of the 5V will appear on the microcontroller I/O pin. Secondly it will not be possible to switch the transistor off as a logic high on the base of the transistor is required to switch a PNP transistor off. This logic high will appear as 3.3V from the microcontroller pin. Current will now flow from the 5V supply through the base-emitter junction to the 3.3V I/O pin because it is at 1.7V lower potential difference than the 5V supply. This will keep the transistor permanently on as switching the microcontroller pin to a logic low will switch the PNP transistor on anyway.

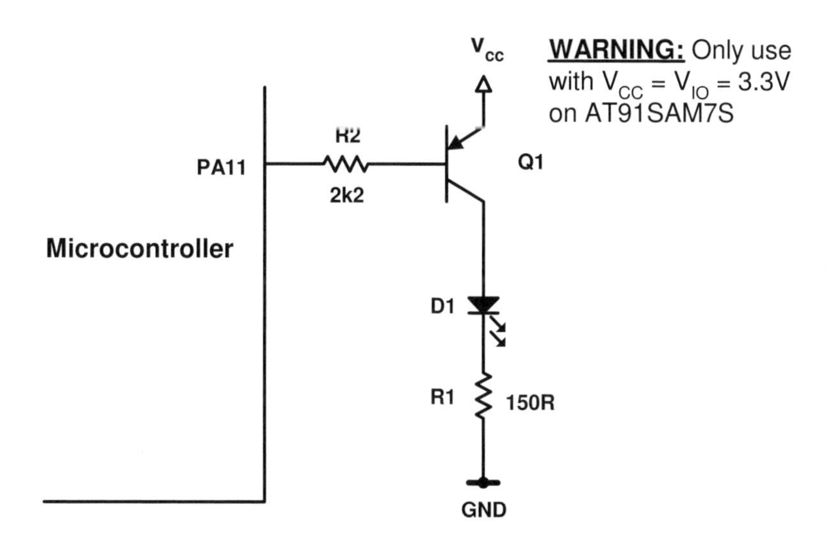

Figure 2.9: Using a PNP Transistor as a Switch

Figure 2.10 shows a solution to switching a load that has a voltage higher than the I/O voltage of the microcontroller, but keeps the transistor that switches the load off at start-up on AT91SAM7S microcontrollers by using a second transistor.

In the figure, transistor Q1 will switch on when the circuit is powered because of the internal pull-up resistor on the microcontrollers pin. R4 is an optional pull-up resistor that does the same job as the internal resistor, but is a stronger pull-up as it has a lower resistance compared to the internal resistor.

With Q1 on, the base of Q2 is pulled low, switching Q2 off and therefore switching power to the load off. When a software program in the microcontroller drives the microcontroller pin low, Q1 is switched off. With Q1 off, the base of Q2 is pulled high by R2, switching Q2 on and therefore switching power onto the load that Q2 is driving.

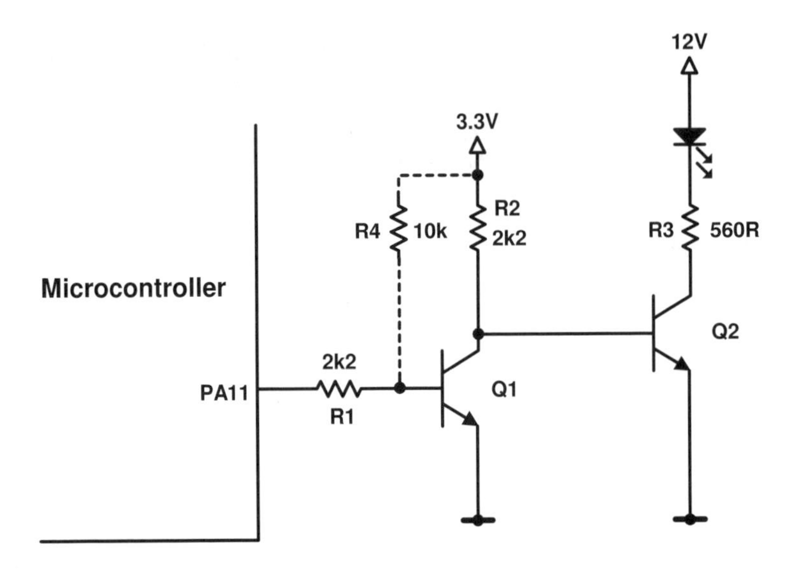

Figure 2.10: Using Two NPN Transistors to Keep the Load Disconnected at Power-up

2.2.2 Field Effect Transistors (FET)

The most common FET transistors used are MOSFETs (Metal Oxide Semiconductor Field Effect Transistors) and more specifically **enhancement mode MOSFETs**. These will be discussed in this section. Another type of FET is the JFET (Junction Field Effect Transistor).

2.2.2.1 MOSFET Interfacing Quick Reference

Figure 2.11 shows how to use an N-channel MOSFET to drive a load. The load is an LED with series resistor again.

The N-channel MOSFETs in **figure 2.11** are performing the same function as the NPN BJTs in **figure 2.4**. The left circuit is switching an LED on that that is operating from the 3.3V supply and the right circuit is switching an LED on that is attached to an external 12V supply.

As with an NPN transistor, the N-channel MOSFET is switched on when a logic high is applied to its gate. The MOSFET in the figure will switch on at power-up when attached to an AT91SAM7S microcontroller because of the internal pull-up resistor. The optional pull-down resistor, R2, can be used to keep the MOSFET off at power-up.

A logic low on the gate of the N-channel MOSFET will switch it off.

Be careful when choosing the MOSFET as many MOSFETs will require too large a gate-source voltage in order to switch on. The gate-source voltage comes from the microcontroller I/O pin and so the requirement will need to be less than 3.3V when used with the AT91SAM7S microcontrollers in order to switch the MOSFET on hard enough.

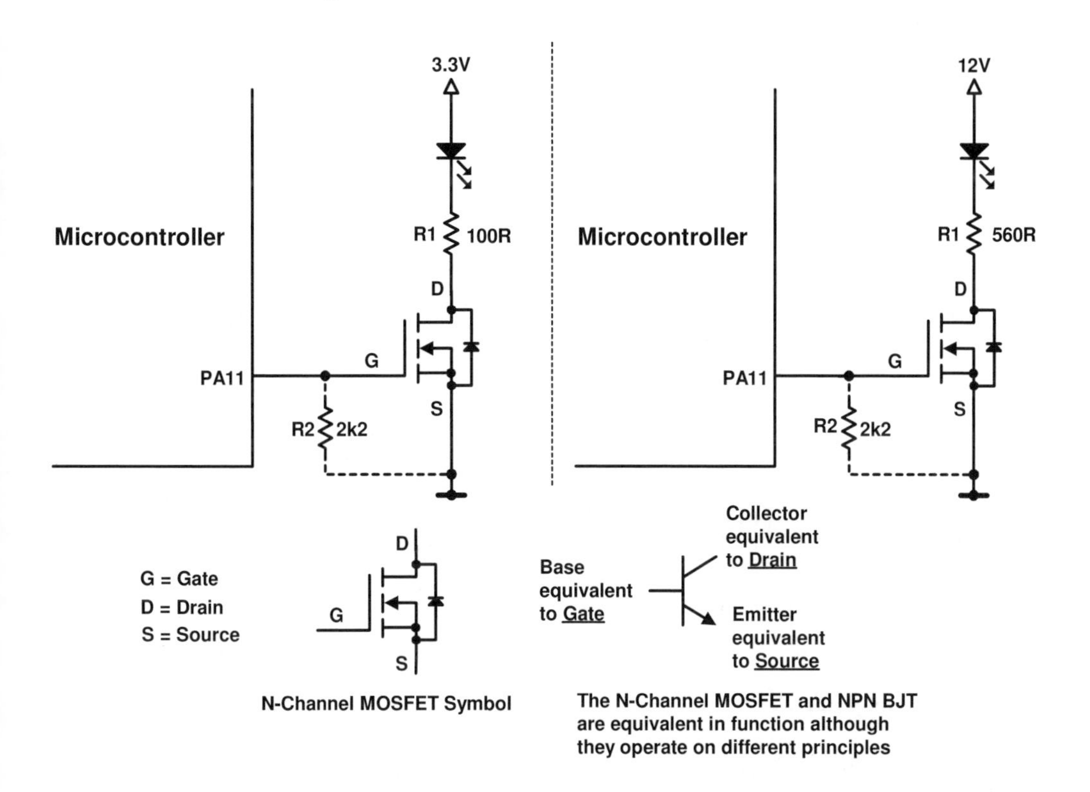

Figure 2.11: N-Channel MOSFET Used as a Switch by a Microcontroller

2.2.2.2 MOSFET Interfacing Explanation and Calculations

How a MOSFET operates as a switch:

While BJTs are current operated devices and require a current to flow through the base-emitter junction in order to switch on, MOSFETs are voltage operated devices and theoretically no current flows into the gate of the MOSFET as the gate is insulated from the rest of the device.

Figure 2.12 shows how an N-channel MOSFET operates as a switch to switch an LED on. The left side of the figure shows that 0V is applied to the gate of the MOSFET (①) which keeps it off, i.e. no current flows from the drain to the source terminal (②).

The right side of the figure shows that when 3.3V is applied to the gate of the MOSFET (①), it switches on, i.e. current flows from the drain to source terminal (②), switching the LED on.

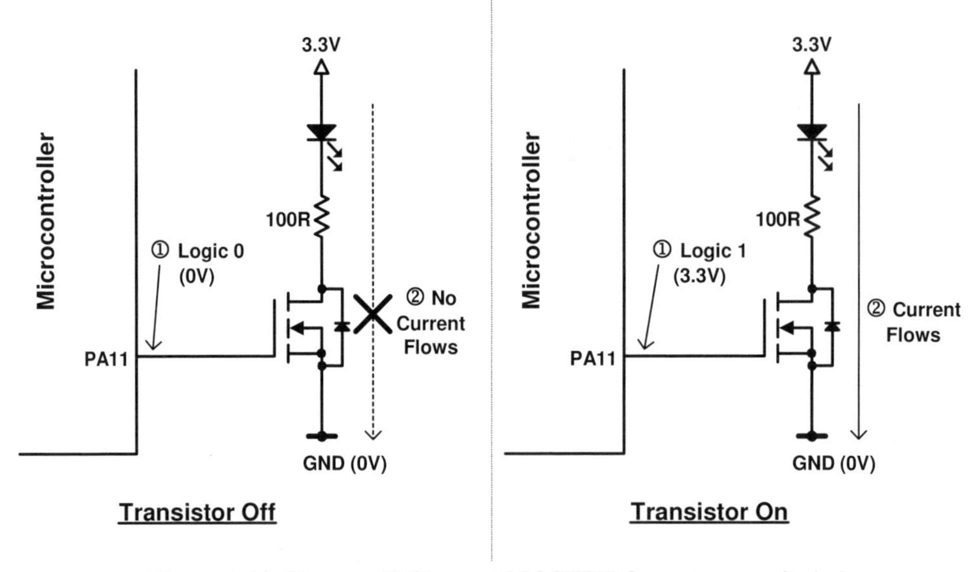

Figure 2.12: How an N-Channel MOSFET Operates as a Switch

MOSFET calculations:

Switching a MOSFET on requires a voltage between the gate and source terminals (V_{GS}). It is important to choose a MOSFET device that can switch on with a low enough V_{GS} so that it can be driven by a microcontroller operating from 3.3V. There are MOSFETs available that are known as "Logic Level" MOSFETs. These devices will switch on with a low V_{GS}. It is important that the gate threshold voltage quoted in the datasheet is well below the microcontroller I/O pin voltage to ensure that the MOSFET gets switched on as hard as possible.

A drain resistor may be included in the circuits of **figure 2.11** and **2.12** to suppress transient voltages that occur when the MOSFET is switched on and off. The drain resistor becomes important when switching the MOSFET at high speeds, for example when using Pulse Width Modulation (PWM) and high power MOSFETs. If desired a gate resistor of between 10Ω and 1kΩ can be included in the low power circuits shown.

Keeping the MOSFET off at power-up or reset:

A pull-down resistor can be used to keep the MOSFET off at power-up. Alternatively, a P-channel MOSFET can be used to keep power to a load off when the AT91SAM7S microcontroller is powered on as shown in **figure 2.13**. In this circuit, writing logic 1 to the port pin will switch the MOSFET off and writing a 0 to the port pin will switch it on. Note that the P-channel MOSFET must only be used to switch a voltage that is the same as or less than the microcontrollers I/O voltage.

If 5V, for example, was used to power the LED in **figure 2.13**, then when 3.3V is applied to the gate of the MOSFET by the port pin to switch it off, there will be a 1.7V difference between the gate and source pins of the MOSFET (5V – 3.3V). This is not a high enough voltage to switch the MOSFET off hard. The BSS84 MOSFET device used to test this circuit has a typical gate-source threshold voltage of 1.7V meaning that this device will only be partly conducting. The result is that when the microcontroller tries to switch the MOSFET off, the LED will be glowing and when the microcontroller switches it on, the LED will glow more brightly.

Figure 2.10 can be used to keep the MOSFET off at power up with a V_{CC} higher than the V_{IO} of the microcontroller. N-channel MOSFETs can be substituted into the circuit for one or both of the NPN transistors.

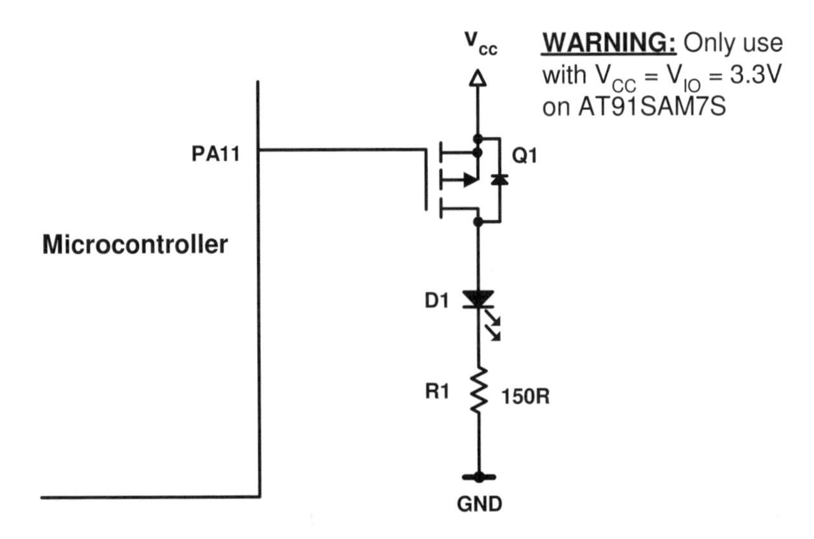

Figure 2.13: A P-Channel MOSFET used as a Switch to Drive an LED

Examples of MOSFET devices:

Figure 2.14 shows examples of N and P-channel MOSFET devices. The examples shown are surface mount devices that are housed in SOT-23 packages. The N-channel device is classed as a "Logic Level" device in the datasheet, but both devices are suitable for use in logic level circuits that use voltages as low as 3.3V.

The N-channel device (BSS138) is capable of switching up to 220mA and the P-channel device (BSS84) is capable of switching up to 130mA.

This section has covered low power MOSFETS to show the principle of operation and application in logic circuits. Other MOSFETs exist that are capable of switching much higher currents and are known as Power MOSFETS. These will be discussed later in this chapter.

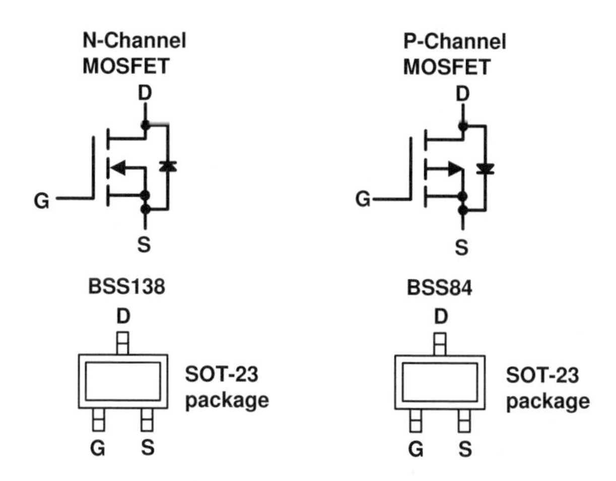

Figure 2.14: Examples of N and P-Channel Logic Level SMT MOSFETs

2.2.3 A Comparison of BJT and FET Transistors

Table 2.4 shows a comparison between BJT and MOSFET transistors.

BJT	MOSFET
Not static sensitive	Static sensitive
Switching times slower than MOSFET (Only significant at high speed operation)	Switching times faster than BJT
Current controlled device (base current) – Uses up current to switch on	Voltage controller device (gate-source voltage) – no current required to switch on
Subject to thermal runaway in extreme conditions – as the device heats up, it conducts more (negative temperature coefficient), causing it to heat up more, etc. eventually destroying the device.	Have a positive temperature coefficient that stops thermal runaway. The resistance between drain and source increases with temperature.

Table 2.4: Comparison of BJT and MOSFET

The choice between BJT and MOSFET for low power applications may depend on the application. For example if designing battery operated equipment a MOSFET will be preferable as it does not need to use up any current to switch on.

For industrial applications, in some instances a BJT may be preferred as it is not static sensitive and so extra circuitry will not be necessary in order to protect the device from high voltage static electricity.

Comparison of a NPN BJT and N-Channel MOSFET Switch Circuit:

Figure 2.15 shows two circuits that were used to compare a BJT and MOSFET by using an AT91SAM7S microcontroller to switch a load of three LEDs on that operate from a 5V supply. The calculated and measured results are shown below the figure.

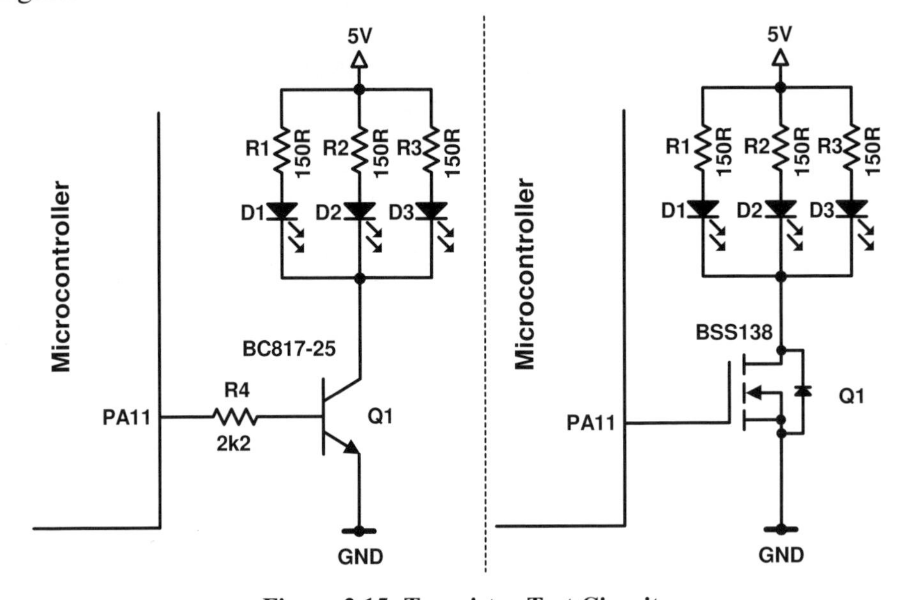

Figure 2.15: Transistor Test Circuits

BJT BC817-25 – when switched on		
	Calculated	**Measured**
Collector Current (I_C)	62mA	56.8mA
Base Current (I_B)	1.18mA	1.1mA
Collector-Emitter Voltage (V_{CE})	≈ 0.2V	0.07V

MOSFET BSS138 – when switched on		
	Calculated	**Measured**
Drain Current (I_D)	66mA	56.7mA
Gate Current (I_G)	≈ 0mA	0mA
Drain-Source Voltage (V_{DS})	0.13V	0.08V

Driving loads that require higher currents:

This section covered only the operating principle of using a transistor as a switch. Later in the chapter, power transistors are used to switch bigger currents.

2.3 Interfacing Optocouplers

Optocouplers (also known as *opto-isolators* or *photocouplers*) can be used with microcontrollers to electrically isolate the microcontroller from the load being driven. This protects the microcontroller from any high voltages that may occur on the load. They can also be used to isolate communication channels such as an RS-485 port, also to protect the microcontroller.

The optocoupler consists of an LED and a photo-transistor in a single I.C. package. To switch a load on, the LED connections of the optocoupler are connected to the microcontroller. When the microcontroller switches the LED on, the photo-transistor which is connected to an external circuit conducts. The result is a completely isolated switching action because only light is passed between the two circuits. **Figure 2.16** shows a microcontroller driving an optocoupler.

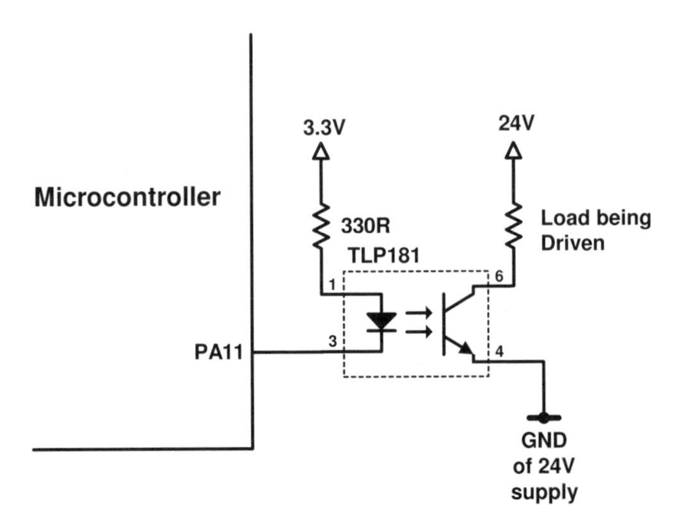

Figure 2.16: A Microcontroller Interfaced to an Optocoupler

Always check the datasheet to see how much current the output of the optocoupler can switch. It may be necessary to use the optocoupler to drive a transistor that can handle the current drawn by the load.

Optocoupler examples for use with 3.3V systems:

TLP181 – SMT optocoupler from Toshiba, transistor output, 3.3V or greater operation

HCPL-060L – SMT/DIP optocoupler from Avago Technologies, logic gate output, 3.3V/5V operation

FOD060L – SMT/DIP optocoupler from Fairchild Semiconductor, compatible with HCPL-060L, but only operates from 3.3V.

2.4 Power Transistors, Relays, Solenoids and Motors

2.4.1 Power Transistors

The transistors covered in section **2.2** of this chapter switched relatively low currents. If we want to switch higher currents we need to use a power transistor.

2.4.1.1 Higher Power BJTs

A medium power transistor, such as the BD139, can have a gain as low as 25. This transistor can switch an absolute maximum of 1.5A. If we wanted to switch only 0.8A, we would need a base current of 0.8A ÷ 25 = 32mA. Adding our 30% margin to make sure that the transistor switches on hard in all conditions: 32mA × 1.3 = 41.6mA. Clearly we cannot switch this transistor on directly from a microcontroller pin.

Figure 2.17 shows a solution to the problem – use a lower power transistor to switch the high power transistor on.

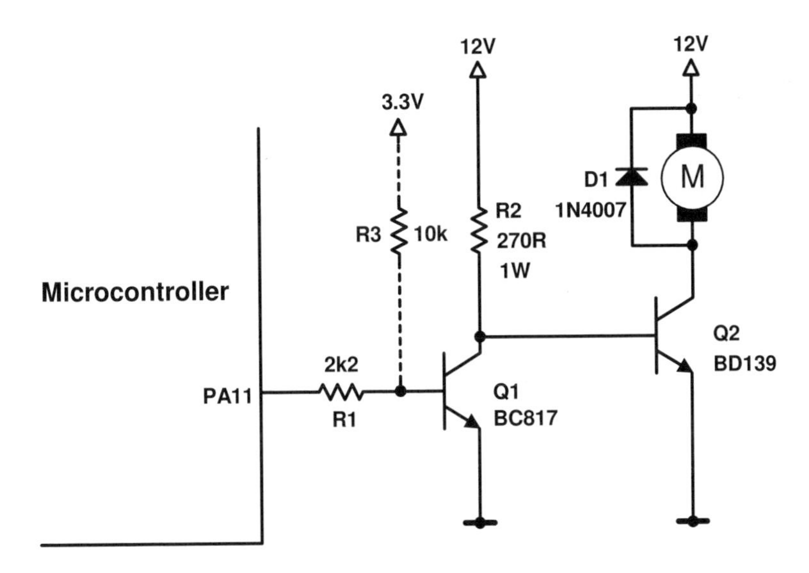

Figure 2.17: Switching a Power Transistor on that Drives a Motor

When the circuit in **figure 2.17** is attached to an AT91SAM7S microcontroller, Q1 will switch on at power-up. The optional pull-up resistor R3 will also help to keep Q1 on. With Q1 on, no current will flow through the base of Q2 as the current will be diverted through Q1 to ground. Q2 will therefore be switched off and the motor attached to it will not run. When Q1 is switched off by the microcontroller, current will flow through R2 into the base of Q2, switching it on. The disadvantage of this circuit is that power is always wasted in R2, as there is always current flowing through it. R2 also needs to be a 1 Watt resistor as more than 0.5 Watts of power is dissipated in it. Note that the diode D1 is necessary in order to protect Q2 from back EMF that is generated by the motor as the motor is an inductive load.

Figure 2.18 shows a similar circuit to **figure 2.17**, but it will switch the motor on at power-up, however power will only be dissipated in R2 when the transistors are switched on. In this circuit, when Q1 is switched on, it switches Q2 on. The figure includes some currents that were measured on a test circuit.

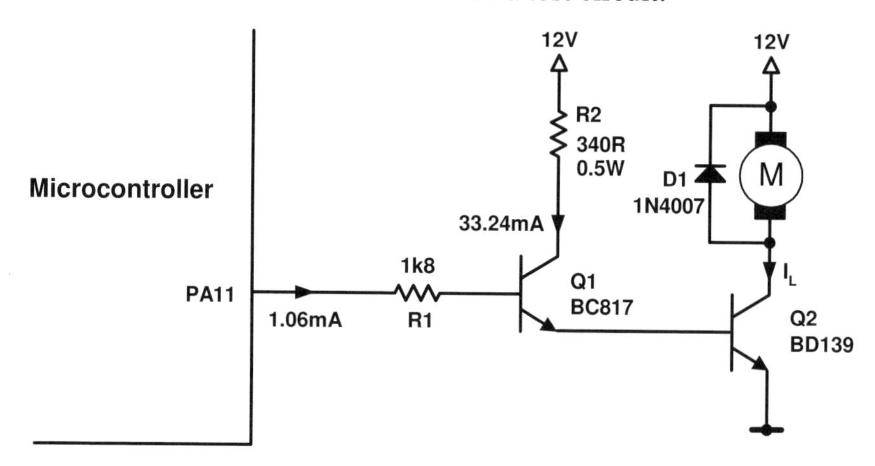

Figure 2.18: Switching a Power Transistor On Alternate Circuit

Table 2.5 shows some measurements taken. I_L is the load current of the load that Q2 is switching i.e. the current that the motor is drawing.

I_L	V_{CE} (Q2)	P (Q2) - Calculated
200mA	0.15V	54mW
300mA	0.21V	87mW
400mA	0.25V	124mW
500mA	0.32V	184mW
600mA	0.37V	246mW
700mA	0.40V	304mW
800mA	0.50V	424mW

Table 2.5: The Effect that an Increasing Load current has on the Transistor – Actual Measurements Taken on the Circuit of Figure 2.18

The table shows that if the base current of a transistor (Q2) stays the same and a bigger current is drawn by the load, V_{CE} will increase. This increase in V_{CE} can be thought of as the transistor trying to switch more off.

The voltage drop across R1 is now calculated as 3.3V – 0.7 – 0.7 = 1.9V. This is because current must flow through the base emitter junction of Q1 and Q2 in order to switch Q1 on.

The power dissipated in a transistor is calculated with the following formula:

$\mathbf{P = V_{CE} \times I_C + V_{BE} \times I_B}$
(In **figure 2.18**, $I_C = I_L$)

In **figure 2.18** the base current of Q2 (I_B) is the sum of the base and collector currents of Q1. I.e. for the measured values in the figure, I_B(Q2) = 33.24mA + 1.06mA = 34.3mA.

To calculate the power dissipated in Q2 for a current of 400mA from **table 2.5**:

$$P = V_{CE} \times I_C + V_{BE} \times I_B$$
$$= 0.25V \times 400mA + 0.7V \times 34.3mA$$
$$= 124mW$$

With the motor drawing 800mA, the transistor dissipates 424mW of power. Under these conditions, the transistor is warm to the touch, but not excessively so.

Increasing the base current of Q2 from 33.24mA to over 80mA with a load current of 800mA decreased V_{CE} to 0.4V. So by more than doubling the base current of the transistor, V_{CE} only decreased by 0.1V.

I personally don't like either of these circuits as the power wastage in R2 seems so unnecessary. A better solution is to use a power MOSFET.

2.4.1.2 Power MOSFETs

Figure 2.19 shows how to use a power MOSFET as a switch. The MOSFET device used in this example is an **IRF540**. This device has a maximum threshold voltage of 4V. The device is really only switched on hard enough with a gate-source voltage (V_{GS}) of 10V.

In the figure, Q1 will switch on at power-up because of the internal pull-up in the AT91SAM7S and the external R3 if connected. This will pull the drain terminal of the MOSFET to ground, switching it off. When the microcontroller switches Q1 off, the gate of the Q2 is pulled up to 12V by R2, switching Q2 on.

Again this is a simple on/off switch and does not control the speed of the motor connected to it. To control the motor speed, see the **chapter 6 – DC Motors and Stepper Motors**.

Comparing the circuit of **figure 2.19** with that of **figure 2.18**, we can note that resistor R2 now draws only about 1.2mA and dissipates only 14.4mW. The on resistance of the MOSFET is also much lower, resulting in much less power dissipation. With 800mA drawn by the load, V_{DS} is only 0.07V, thus the power dissipated in the MOSFET is now only 56mW compared to 424mW dissipated by the BJT. The MOSFET is cool to the touch when operating at this current.

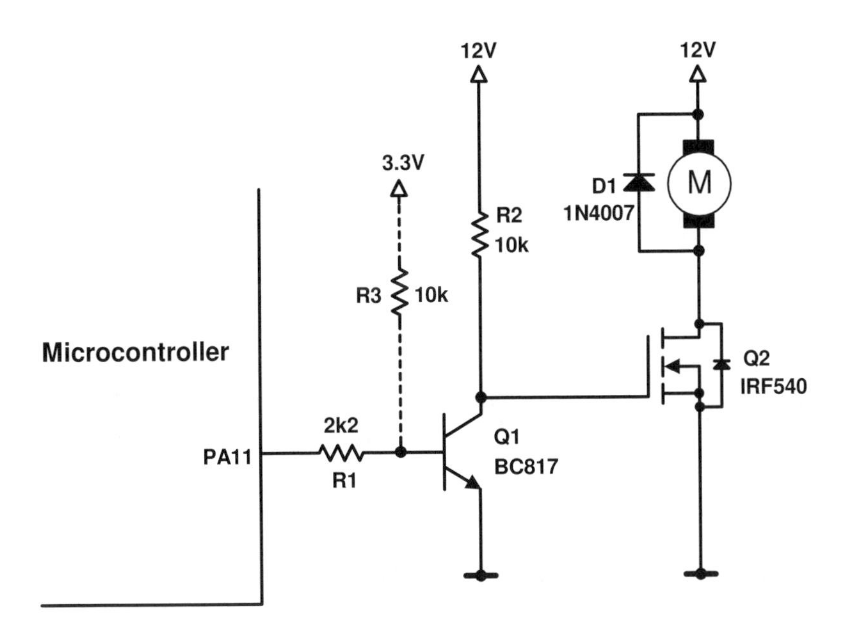

Figure 2.19: Using a Power MOSFET as a Switch

The absolute maximum gate-source voltage for the IRF540 is 20V. If switching a higher voltage circuit, we need to use a voltage divider to reduce the voltage across the gate-source junction of the MOSFET. **Figure 2.20** shows how this is done. The circuit also incorporates an optocoupler for better isolation of the higher voltage circuit from the microcontroller circuitry.

When the microcontroller is powered up, the LED in the optocoupler is off. The transistor in the optocoupler is therefore like an open circuit, resulting in the gate of Q1 being pulled down to ground by R3. When the microcontroller switches the optocoupler on, R2 and R3 act as a voltage divider, dividing the 24V supply voltage in half resulting in a gate-source voltage of 12V that switches Q1 on.

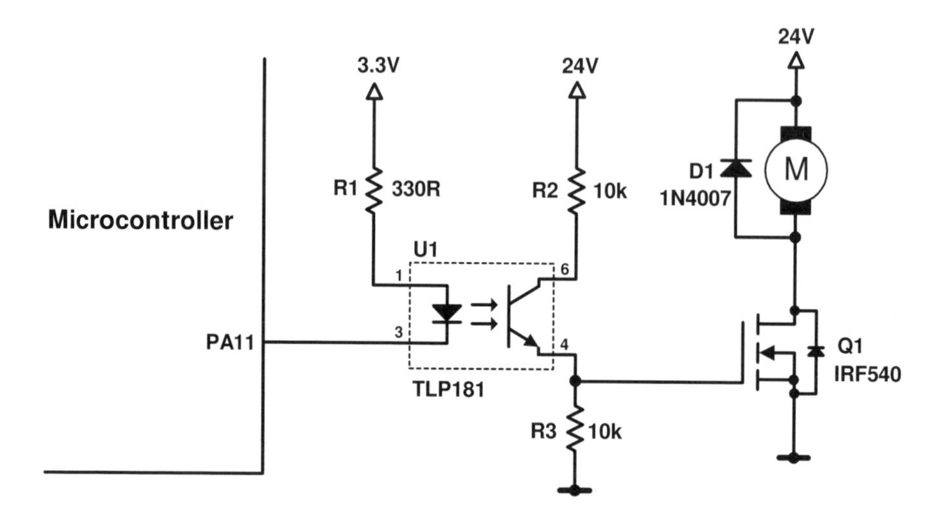

Figure 2.20: Interfacing a MOSFET to a Microcontroller using an Optocoupler

2.4.2 Relays and Solenoids

Relays and solenoids (and motors) are inductive devices so they must have a protection diode across the inductive load to protect the transistor that drives them from back EMF. **Figure 2.21** shows how to drive a 12V d.c. relay. The contacts of the relay can be used to switch an a.c. or d.c. load. A solenoid can be driven by the same circuit. The circuit shows an optional LED that can be used to indicate when the relay or solenoid is active. A pull-down resistor on the base of Q1, or a circuit like **figure 2.10** can be used if the relay or solenoid must be kept off when the AT91SAM7S microcontroller is powered up.

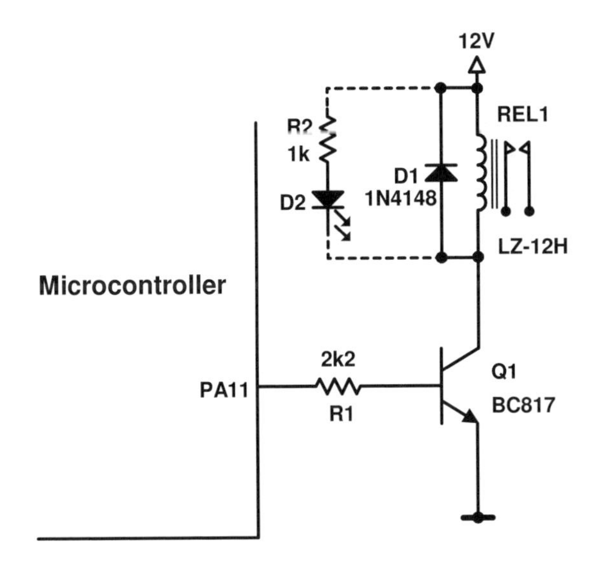

Figure 2.21: A Microcontroller Driving a Relay

In the circuit of **figure 2.21**, the relay coil drew 31.8mA (Takamisawa 12V relay). The LED will draw about 10mA.

The relay switch contacts can be used to switch an a.c. or d.c. circuit that is connected to it.

Figure 2.22 shows a circuit that will drive the relay using an optocoupler. The relay will remain off at power-up when attached to an AT91SAM7S microcontroller. The relay is not driven directly by the optocoupler as the optocoupler can't deliver enough current to switch the relay. Instead, Q1 is switched on by the optocoupler to drive the relay.

R3 is a pull-down resistor that keeps Q1 off when the optocoupler is off.

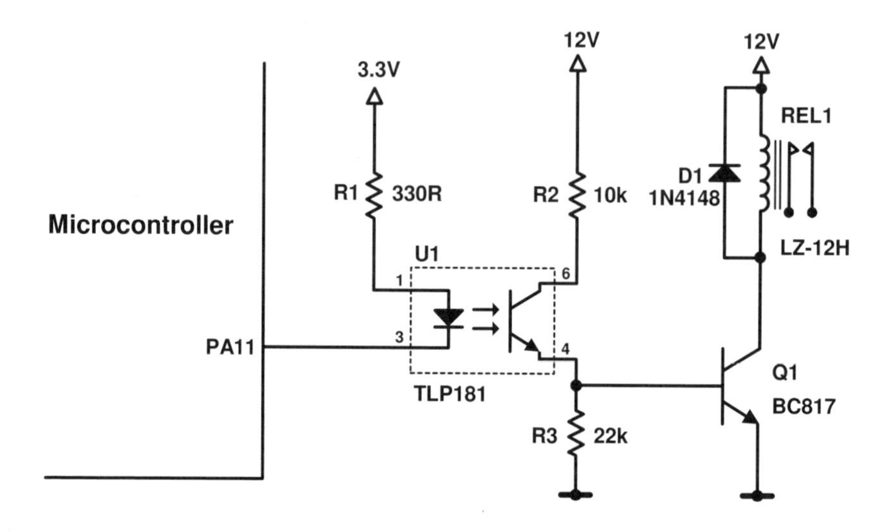

Figure 2.22: Optocoupled Relay Driver

2.5 Software for Driving Digital Outputs

2.5.1. What the Software Must Do

In order to control any of the circuits in this chapter with software, we need to set up the microcontroller's I/O pin that drives the circuit as an output and then be able to toggle the state of that pin (change it between logic 0 and logic 1 states).

The software to set up and use the I/O pins has to be specific to our chosen microcontroller, the AT91SAM7S range. If you want to use a different microcontroller, you will need to refer to the datasheet and software examples for the different microcontroller.

On the AT91SAM7S

On AT91SAM7S microcontrollers, the **Parallel Input/Output Controller (PIO)** is used to control the I/O pins. Each I/O pin can be individually setup as an input or output and toggled on and off. The PIO is an internal peripheral device to the AT91SAM7S.

The PIO controller and most other AT91SAM7S internal peripheral devices have a clock input that must be enabled for the device to operate properly. The clock for each device is normally disabled at power-up and must be enabled in the **Power Management Controller (PMC)**. For simple output operation, this clock does not need to be enabled and is therefore not enabled in the software examples of this chapter.

At power-up all I/O pins on the AT91SAM7S are enabled as inputs with internal pull-up resistors enabled. We need to change at least one of these pins to an output and may want to disable the pull-up resistor.

On the AT91SAM7S, the PIO A controller is present. Some other AT91SAM7 microcontrollers have a PIO B controller that controls additional I/O pins. The registers for the AT91SAM7S microcontrollers that control the PIO A controller have been defined in the **`at91sam7s.h`** file with names starting with PIOA. This file has been provided as part of the template files for use with this book.

The information on how to set up the PIO controller is obtained from the AT91SAM7S datasheet, so refer to it when working with the PIO controller.

At the time of writing, the newest version of the datasheet was used:
AT91SAM7S Series Preliminary (757 pages, revision I, updated 12/08)

The web link to this datasheet is:
www.atmel.com/dyn/resources/prod_documents/doc6175.pdf

Any peripheral device on the AT91SAM7S, such as the PIO controller is controlled by writing to the set of registers associated with the device. The register definitions for the PIO controller are provided at the end of chapter 27 in the AT91SAM7S datasheet.

The AT91SAM7S has 32 I/O pins PA0 to PA31. Each bit in any of the 32 bit PIO registers corresponds to one of the port pins. I.e. manipulating bit 3 in one of the PIO registers affects the corresponding PA3 pin. In the datasheet, the bits in each register are labelled P0 to P31, corresponding to I/O pins PA0 to PA31. The PIO registers are labelled starting with PIO not PIOA in the datasheet.

2.5.2 Example Software Programs

Three example programs have been provided to drive output pins; the first two are set up to run from internal SRAM memory:

digital_output_RAM_basic – shows basic operation of single I/O pin as output
digital_output_RAM – switches more than one output pin and defines some pin names
digital_output_Flash – same as **digital_output_RAM**, but compiles to run from Flash

The code listing of **digital_output_RAM_basic** is shown below:

```
#include "at91sam7s.h"

void Delay(void);

int main(void)
{
  /* configure PIO line PA11 as output */
  PIOA_PER = 0x00000800;    /* enable uC pin as PIO */

  PIOA_PUDR = 0x00000800;   /* disable the pull-up resistor */

  PIOA_OER = 0x00000800;    /* enable uC pin as output */

   while (1) {
     PIOA_SODR = 0x00000800; /* set PA11 pin state to high */
      Delay();
      PIOA_CODR = 0x00000800; /* set PA11 pin state to low  */
      Delay();
   }
}

void Delay(void)
{
    int del = 2000000;

    while (del--);
}
```

Contains AT91SAM7S register definitions

In the **digital_output_RAM_basic** program, the I/O pin PA11 is set up as follows:

```
PIOA_PER = 0x00000800;
```
This line enables the PIO controller to control PA11 as an I/O pin by setting bit 11 (P11) in the PIOA_PER. PER = PIO Enable Register.

```
PIOA_PUDR = 0x00000800;
```
This line disables the internal pull-up resistor connected to PA11. If the pull-up resistor must be enabled, this line can be left out or can be explicitly enabled by setting bit 11 in the PIOA_PUER. PUDR = Pull Up Disable Register, PUER = Pull Up Enable Register.

```
PIOA_OER = 0x00000800;
```
Enables PA11 as an output by setting bit 11 in PIOA_OER. OER = Output Enable Register.

At this stage I/O pin PA11 has been set up as an output with the internal pull-up resistor disabled. The code in the **while** loop now switches PA11 high and low with a short delay in-between:

```
PIOA_SODR = 0x00000800;
```
This sets the I/O pin PA11, i.e. it is switched to a logic high level (logic 1). SODR = Set Output Data Register.

```
PIOA_CODR = 0x00000800;
```
This clears I/O pin PA11, i.e. it is switched to a logic low level (logic 0). CODR = Clear Output Data Register.

Which of these lines of code switches the device attached to it on depends on how the device that is being switched is connected. In **figure 2.1**, the current sourcing LED will be switched on by logic 1 (SODR), and off by logic 0 (CODR). The current sinking LED circuit will work the opposite way.
Note that when writing to any of the PIO registers to set up or control PA11, bit 11 of each register is set. Most of the registers have a set and clear pair. Setting a bit in a "set" register performs a certain function and setting a bit in a "clear" register disables the function. With this arrangement an individual bit that controls a function can be switched on or off without worrying about affecting other bits in the register. For example each time that we write 0x00000800 to a register, we are setting bit 11, but are clearing all the other bits in the register.

Because writing a 1 to the register is the only thing that has an effect on the register, writing 0 to the other bits does nothing. We therefore do not have to first read the contents of the register, manipulate a single bit and then write the contents back to the register.

The PIO controller can be set up to write a whole byte or even all 32 bits at the same time with a single write. This will be covered in **chapter 5 – LCD and Seven Segment Displays** where it is desirable to write a byte or nibble at a time instead of setting and clearing individual bits.

2.5.3 Alternate Program

The **digital_output_RAM** and **digital_output_Flash** programs have been provided to show a different way of setting individual bits in a register by giving them a name using **#define**:

```
#define PA0         (1 << 0)
#define PA1         (1 << 1)
#define PA2         (1 << 2)
#define PA3         (1 << 3)
#define PA11        0x00000800
```

Each bit in the program could be defined as PA11 has been in the above listing, but an easier way is to use the notation as shown with PA0 to PA3. 1 is shifted left x number of times, where x is the bit number in the register:

(1 << 0) is the same as 0x00000001
(1 << 1) is the same as 0x00000002
(1 << 2) is the same as 0x00000004
(1 << 3) is the same as 0x00000008

PA11 could be defined as:
(1 << 11)

OUT_PINS is a name used that combines a number of bits using the C 'OR' operator in order to initialise the corresponding pins at the same time:

```
#define OUT_PINS    (PA11 | PA0 | PA2)
```

In this example, the same registers are used to set up all three pins simultaneously:

```
PIOA_PER  = OUT_PINS;     /* enable uC pin as PIO          */
PIOA_PUDR = OUT_PINS;     /* disable the pull-up resistor */
PIOA_OER  = OUT_PINS;     /* enable uC pin as output      */
```

This is the same as writing:

```
PIOA_PER  = 0x00000805;
```

etc…

Now I/O pins can be enabled or disabled as outputs at the top of the C source file by adding or removing pin names to OUT_PINS.

The same pin definitions are used to toggle the enabled pins:

```
/* toggle the outputs */
while (1) {
    PIOA_SODR = PA11;
    PIOA_CODR = PA0;
    PIOA_SODR = PA2;
    Delay();
    PIOA_CODR = PA11;
    PIOA_SODR = PA0;
    PIOA_CODR = PA2;
    Delay();
}
```

2.5.4 Other Considerations

It may not be desirable to have a transistor or LED that is attached to a microcontroller pin switch on immediately after the PIO controller has been set up. This will occur when a circuit such as **figure 2.9** is used.

To prevent the transistor from switching on right after initialisation, first set the pin using the PIOA_SODR (the pin logic level will default to 0 after reset and after the pin has been enabled as an output). Now when the PIO controller is initialised, the pin state will remain high instead of immediately switching low:

```
/* configure PIO line PA11 as output */
PIOA_SODR = 0x00000800;    /* first set pin to logic 1 */

PIOA_PER = 0x00000800;     /* enable uC pin as PIO */
PIOA_PUDR = 0x00000800;    /* disable the pull-up resistor */
PIOA_OER = 0x00000800;     /* enable uC pin as output */

Rest of program...
```

3. Timers and Interrupts

In this chapter we look at the on-chip timers of the AT91SAM7S as well as how to set up and use interrupts in a C program. Both of these topics will be very useful when writing further programs to control hardware.

3.1 Timers

The AT91SAM7S has four types of timers:

1) 1 × **Real-time Timer (RTT)** – not accurate, runs from internal RC oscillator.

2) 1 × **Periodic Interval Timer (PIT)** – accurate, intended for operating system's scheduler interrupt.

3) 1 × **Watchdog Timer (WDT)** – for resetting the microcontroller should a system lock-up occur in software.

4) 3 × **Timer Counter (TC) Channels** – multi-function timer/counter channels for timing, counting, pulse generation and more.

3.1.1 Real-time Timer (RTT)

The RTT starts running when the AT91SAM7S microcontroller is powered up. In this default state it increments a count value every second. The **Real-time_timer_RAM** program reads the time value from the RTT_VR for display on a terminal emulator:

```
#include <stdio.h>
#include "at91sam7s.h"
#include "dbgu.h"
```

```
int main(void)
{
    char str[50];

    DBGUInit(DBGU_115200);

    DBGUTxMsg("\r\nTesting Real Time Timer (RTT)\r\n");

    while (1) {
        sprintf(str, "%10d\r", RTT_VR);
        DBGUTxMsg(str);
    }
}
```

The RTT has only four registers and is pretty straight forward to use. The mode register (RTT_MR) is used to set the prescaler which sets the time period of the timer. It also contains two bits for controlling interrupts and one for restarting the timer.

The alarm register (RTT_AR) contains the value to compare with the RTT time in order to trigger an interrupt. The timer value register (RTT_VR) contains the current time value of the RTT. The status register (RTT_SR) contains only two bits that indicate if the alarm register value was reached and if the timer has been incremented since the last read of this register.

The Slow Clock (SLCK)
The clock source for the RTT is the slow clock. The slow clock is an RC oscillator that is embedded in the AT91SAM7S microcontroller. The frequency of the slow clock is **32.768kHz**, but is not accurate and can vary between **22kHz** and **42kHz**. A slow clock frequency of 32768Hz will cause the RTT to count in 1 second intervals when the prescaler (RTPRES bit field in RTT_MR) is set to 0x8000 because 32768 = 8000H, so to count up to 8000H (32768) will take one second.

3.1.2 Periodic Interval Timer (PIT)

The PIT is intended to be used to generate an interrupt for an operating system's scheduler. If you are not running an operating system on your embedded system, you can use the PIT for any task that you desire. The **`Periodic_Interval_Timer_RAM`** program starts the PIT and displays the two count values in the PIT_PIIR on a terminal emulator:

```
#include <stdio.h>
#include "at91sam7s.h"
#include "dbgu.h"

int main(void)
{
    char str[50];
    unsigned int count_val;

    DBGUInit(DBGU_115200);

    DBGUTxMsg("\r\nTesting PIT\r\n\r\n");

    PIT_MR = 0x010FFFFF;

    while (1) {
        count_val = PIT_PIIR;
        sprintf(str, "CPIV = %7d\t\tPICNT = %4d\r",
                count_val & 0x000FFFFF, count_val >> 20);
        DBGUTxMsg(str);
    }
}
```

In the program, the value written to the PIT_MR (mode register) starts the PIT and sets it up to count to the maximum period before incrementing the count. The **`while(1)`** loop is used to display the values of the two bit fields in the PIT_PIIR (Periodic Interval Timer Image Register). The CPIV (Current Periodic Interval Value) field contains the number of master clock (MCK) cycles ÷ 16. When the CPIV field reaches the value that was written in the PIV field of the PIT_MR register, the PICNT (Periodic Interval Counter) field is incremented.

The PIT contains only four registers:

PIT_MR (Periodic Interval Timer Mode Register) – for setting up the timer period, enabling the timer and enabling the timer interrupt.

PIT_SR (Periodic Interval Timer Status Register) – contains one bit (PITS) that indicates if the timer has ticked.

PIT_PIVR (Periodic Interval Timer Value Register) – contains the timer values. Reading this register resets the PITS bit in the PIT_SR.

PIT_PIIR (Periodic Interval Timer Image Register) – identical to the PIT_PIVR register, but reading this register has no effect on any other register bits.

The Master Clock (MCK)
The clock source for the PIT is MCK ÷ 16. The master clock can be generated by one of several clock sources, but in the example programs used in this book it is generated by the PLLCK. This is setup in the assembly language file **`startup.s`** supplied with the source code. It is assumed that an **18.432MHz** crystal is connected to the AT91SAM7S oscillator pins.

The master clock is generated as follows:

1) The main oscillator generates a MAINCK of 18.432MHz because of the **18.432MHz** crystal connected to XIN and XOUT oscillator pins of the AT91SAM7S.

2) The MAINCK is fed into a divider and PLL.

3) The divider is set up by the DIV field in the CKGR_PLLR (PMC_PLLR) of the Power Management Controller (PMC). In the example code, DIV = 5, so the oscillator frequency is now 18.432MHz ÷ 5 = 3.6864MHz.

4) In the PLL, the MUL field in the CKGR_PLLR (PMC_PLLR) multiplies the frequency by MUL + 1 times. In the example code, MUL = 25, so PLLCK is now 3.6864MHz × (25 + 1) = 95.8464MHz.

5) The PLLCK is then fed into the master clock prescaler to generate the master clock (MCK). The prescaler value is set in the PRES field of the PMC_MCKR. In the example code, the prescaler is set to 2, so the MCK frequency is now 95.8464 ÷ 2 = **47.9232MHz**.

With MCK = 47.9232MHz, the input frequency to the PIT is 47.9232MHz ÷ 16 = **2.9952MHz** because the MCK is passed through a prescaler that divides it by 16. This gives a clock period of 1 ÷ 2.9952MHz = **333.8675nS**.

To generate a 1mS tick, the value to write to the PIV field in PIT_MR will be 1mS ÷ 333.8675nS – 1 = 2995.2 – 1 = **2994**.

3.1.3 Watchdog Timer (WDT)

When enabled, the watchdog timer will reset the microcontroller should the software running on the microcontroller "hang up" or crash. The embedded program running on the microcontroller must periodically reset the watchdog timer. If the software hangs up at some point in the program, the watchdog timer will not be reset by the software and will time out. When it has timed out, it will reset the microcontroller.

The watchdog timer on the AT91SAM7S microcontroller is enabled automatically when the microcontroller is powered up. The timer inside this watchdog timer is set up by default to count 16 seconds and when the 16 seconds are up, it resets the microcontroller.

On the AT91SAM7S, the watchdog timer can be set to count a maximum of 16 seconds before timing out. The watchdog timer operates from the slow clock and the maximum value of 16 seconds can be obtained if it is running at 32.768kHz. Remember that the slow clock is not accurate.

At power-up (reset), the watchdog timer is enabled. The template files used with this book disable the watchdog in the assembly language file. This is why the programs that you have written using the template files do not cause the microcontroller to be reset every 16 seconds.

The register that allows the setup or disabling of the watchdog timer can only be written to once after reset. We must therefore either configure it in the assembly language file (set it up there instead of disabling it), or remove the code that disables it so that we can set it up in our C source file. If you have been using SAM-BA to load code into the AT91SAM7S microcontroller's SRAM memory via the USB port, the SAM-BA bootloader code will have already written to the watchdog setup register, so when your program writes to it, it will have no effect. The example program that follows must be loaded into Flash memory, so that it will run at power-up and the program will be the first thing that writes to the watchdog setup register. There are some bits in this register that will cause the watchdog timer to stop running when the microcontroller is in debug mode. This will stop the microcontroller from being reset by the watchdog timer if you are debugging using an ICE.

In the **Watchdog_Flash** program, the code in the assembly language file (**startup.s**) that disables the watchdog timer is commented out to prevent the watchdog from being disabled and then causing the C program write to the watchdog register to fail. This program must be loaded to the Flash memory of the microcontroller:

```
#include <stdio.h>
#include "at91sam7s.h"
#include "dbgu.h"

#define WDT_RESET_VAL   0xA5000001

void Delay(void);

int main(void)
{
    int count = 0;
    char str[50];

    /* set watchdog timer up to time out after 5 seconds */
    WDT_MR = 0x35002500;

    DBGUInit(DBGU_115200);

    DBGUTxMsg("\r\nProgram starts from beginning...\r\n");
    DBGUTxMsg("Entering loop that restarts watchdog.\r\n\r\n");

    /* watchdog timer is restarted in this loop */
    while (1) {
        Delay();
        sprintf(str, "\rcount = %4d", count);
        DBGUTxMsg(str);
        count++;
        WDT_CR = WDT_RESET_VAL;   /* restart the watchdog timer */
        if (count > 50) {
            count = 0;
            break;
        }
    }

    DBGUTxMsg("\r\n\r\nSimulating a software hang up.\r\n");
    DBGUTxMsg("Expect the microcontroller to be reset...\r\n\r\n");
```

```
    /* watchdog timer is not restarted in this loop */
    while (1) {
        Delay();
        sprintf(str, "\rcount - %4d", count);
        DBGUTxMsg(str);
        count++;
    }
}

void Delay(void)
{
    int count = 1000000;

    /* waste time */
    while (count) {
        count = count - 1;
    }
}
```

At the start of the program, the watchdog timer is set up to time out after 5 seconds by writing to the WDT_MR (Watchdog Timer Mode Register), a message is then sent to the serial port for display on a terminal emulator on the PC. This message is used to see when the watchdog resets the microcontroller as it will only be run once at reset or power-up.

The first **while(1)** loop demonstrates how a program should periodically reset the watchdog timer while displaying a count on the terminal emulator.

The second **while(1)** loop simulates what will happen if the software locks up and cannot reset the watchdog timer. The watchdog timer will reset the microcontroller after 5 seconds and the reset will cause the program to start running from the beginning, displaying the start-up message again.

Note that the count displayed is not in seconds, but only serves to show that the microcontroller is busy with a task.

In the program, the WDT_MR is setup with its WDV (Watchdog Counter Value) and WDD (Watchdog Delta Value) fields set to 0x500 (1280). This is to obtain a 5 second timeout value for the watchdog timer and assumes that the slow clock is running at 32.768kHz. The slow clock is divided by 128 before it is fed to the WDT.

The formula for calculating the timeout register value is:

$$\begin{aligned} WDT_{COUNT} &= \frac{Timeout}{\left(\frac{1}{32768}\right)\times 128} \\ &= \frac{5}{\left(\frac{1}{32768}\right)\times 128} \\ &= 1280 \\ &= 0x500 \end{aligned}$$

Where:
WDT_{COUNT} = count value to load in WDT register (maximum = 0xFFF = 4095)
$Timeout$ = WDT timeout value required - in seconds

3.1.4 Timer Counter (TC)

The Timer Counter can be set up in two main modes, namely **capture mode** – for counting and timing external clock pulses (measuring signals) and **waveform mode** – for timing purposes and wave generation. The **TC0_RAM** program sets up Timer Counter 0 (TC0) in waveform mode to count in milliseconds if the master clock is set up as in the template source files.

```
#include <stdio.h>
#include "at91sam7s.h"
#include "dbgu.h"

void TimeDelay(unsigned int ms_delay);

int main(void)
{
    char str[50];
    unsigned int time = 0;
    /* enable the clock of the TC0 module through the PMC */
    PMC_PCER = 0x00001000;
    /* setup TC0 registers */
    TC0_CMR = 0x0000C004;  /* wave mode, clock = (MCLK / 1024)*/
    TC0_RC = 0x0000002F;         /* RC Register value for 1ms */
    DBGUInit(DBGU_115200);
```

```
    while (1) {
        sprintf(str, "%010d\r", time);
        DBGUTxMsg(str);
        TimeDelay(1000);
        time++;   <------------   Counts up in seconds
    }
}

void TimeDelay(unsigned int ms_delay)
{
    TC0_CCR = 0x00000005;        /* start the timer */
    while (ms_delay) {
        if (TC0_SR & 0x00000010)
            ms_delay--;   <------------   Counts down in milli-seconds
    }
}
```

This program sets up TC0 to count in milliseconds. A function is called which uses the timer to count in milliseconds. The program uses the function to print a count value to the terminal emulator every second.

At the start of the program, the clock to TC0 is enabled in the PMC_PCER register by setting the bit corresponding to its PID – PID12.

The TC0_CMR (Channel Mode Register) sets up TC0 in waveform mode by setting the WAVE bit, sets it up to restart the counter after timing out (CPCTRG bit is set) and sets the TCCLKS bit field to supply it with a clock frequency of MCLK / 1024.

The TC0_RC register sets up TC0 to count in milliseconds. In the **`TimeDelay()`** function, the timer is started by writing to the TC0_CCR (Channel Control Register). The CPCS bit in the TC0_SR (Status Register) is then checked to see if 1ms has passed, if it has the millisecond counter variable **`ms_delay`** is decremented.

3.2 Interrupts

Why interrupts are needed:
A program can be written to **poll** (continually check) a hardware device to see if it is ready to be read from or written to. An example is a serial port: it can be continuously polled to see if data has arrived so that the data can be read and used by the program. This wastes processor time as the program needs to sit in a loop and check for arrival of data. Other hardware devices may also need to be serviced while the program is stuck in a loop. Interrupts can solve this problem.

How interrupts work:
Hardware devices can be set up to use interrupts. When using interrupts, the hardware will interrupt the normal flow of a program so that the program can service the hardware (read from or write to) and then return to the normal program when finished. When an interrupt is asserted, the microcontroller will save the current state of the program that is running, jump to a special function called the "**interrupt service routine (ISR)**" (written by the programmer). The interrupt service routine will run code that services the hardware. When the interrupt service routine returns, the state of the program that was running is restored and the program continues running where it left off before the interrupt.

As an example, if the serial port is set up to use an interrupt when a data byte is received, the serial port will assert the interrupt line. The main program will then stop and the current state of the processor will be saved so that it can be restored after the interrupt has been serviced. The interrupt service routine will then be run. This all happens automatically if interrupts have been set up. The programmer has written the ISR to, for example, save the received byte to an array. The ISR will then return, the state of the processor will be restored and the main program will continue running.
By now you will understand why you must never use an interrupt service routine to reset the watchdog; it may continue to run correctly, thus kicking the watchdog, even if all the main software is stuck.

3.2.1 Interrupts on the ARM7 and AT91SAM7S

The ARM7 core has only two interrupt lines: nIRQ and nFIQ. nIRQ is a normal interrupt request line and nFIQ is a fast interrupt request line. When either of these lines is asserted, the ARM7 core is interrupted.

The AT91SAM7S microcontroller contains a piece of hardware, the **advanced interrupt controller (AIC)**, which controls the two interrupt lines of the ARM7 core. Interrupts can be generated by hardware devices on the AT91SAM7S (such as the timer or serial port) or by an external device connected to one of the AT91SAM7S external pins. When an interrupt is generated, it will notify the AIC, which will provide the address of the interrupt service routine (ISR) as well as assert the interrupt line of the ARM7 core. This will result in the ARM7 running the correct ISR for the corresponding interrupt. The address of each interrupt service routine to be used for each interrupt source is written to the AIC when setting up interrupts at the start of a program. When the AIC receives an interrupt, it will fetch the address of the corresponding ISR. **Figure 3.1** shows a block diagram of the AIC.

The AIC contains an 8-level priority controller which allows the interrupts to be set at different priority levels. This will allow a higher priority interrupt to interrupt a lower priority interrupt.

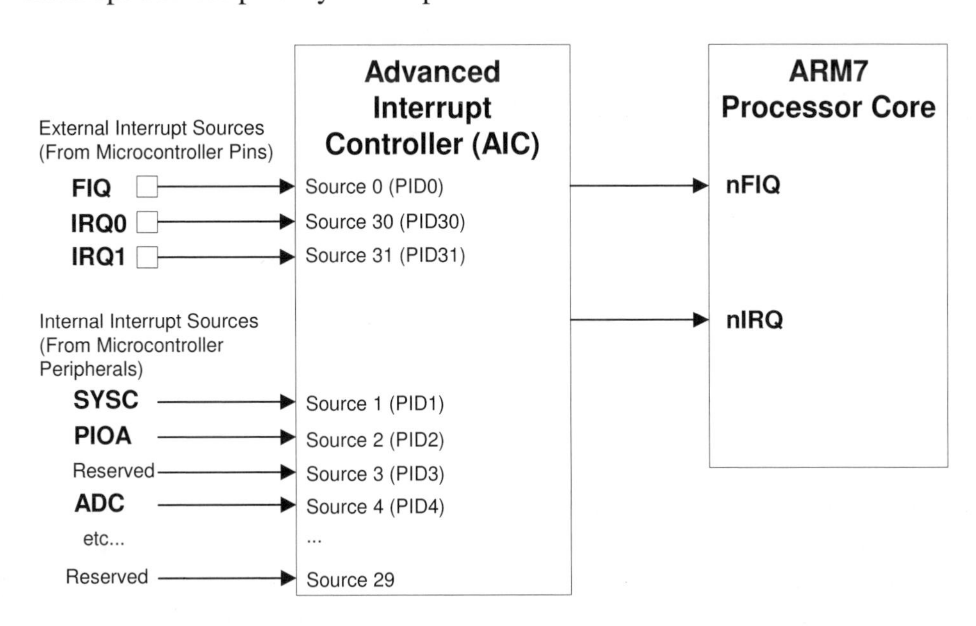

Figure 3.1: The AT91SAM7S Advanced Interrupt Controller (AIC)

Interrupts can be handled in one of two ways by the C program. **The first way** is to use an extension to the C language that is provided by the GCC compiler. This does not allow **nested interrupts**.

Nested interrupts – if an interrupt occurs and then a different higher priority interrupt occurs, the higher priority interrupt will be allowed to interrupt the first interrupt. Lower priority interrupt will be blocked until a higher priority interrupt has finished being serviced.

The second way of handling interrupts is to put the interrupt specific code in the assembly language start-up file. The assembly code will then re-enable interrupts when an interrupt occurs, allowing a higher priority interrupt to interrupt a lower priority interrupt (**nested interrupts**). The interrupt service routine can then be written just like any C function, without any C language extensions by the compiler to handle interrupts.

3.2.2 Non-nested Interrupts using GCC Extension

The example program, **`TC0_Interrupt_nonest_Flash`** shows the first way of setting up interrupts in C. Only part of the code listing of this example is shown here. Refer to the downloaded code for the full listing.

In the **`startup.s`** assembly language file:

```
/*-----------------------------------------------------------------
    Interrupt Vectors
-----------------------------------------------------------------*/
/* exception vectors at address 0 */
Vectors:     LDR   PC, =Reset_Handler   /* reset vector            */
             LDR   PC, =Undef_Handler   /* undefined instruction */
             LDR   PC, =SWI_Handler     /* software interrupt    */
             LDR   PC, =PAbt_Handler    /* prefetch abort        */
             LDR   PC, =DAbt_Handler    /* data abort            */
             NOP                        /* reserved vector       */
             LDR   PC,[PC,#-0xF20]      /* IRQ vector            */
             LDR   PC, =FIQ_Handler     /* FIQ vector            */
```

The second instruction from the bottom of the interrupt vector table (IRQ vector) loads the contents of the AIC_IVR (Interrupt Vector Register) to the PC (Program Counter). The AIC_IVR contains the address of the interrupt service routine that is in the C source code that corresponds to the interrupt that occurred. When this address is loaded into the PC, the ISR will be run.

In **main.c**, the following line of code contains GCC extensions to the C language to generate the correct interrupt code for the ISR. It is placed at the function prototype for the ISR:

```
/* TC0 ISR (interrupt service routine) */
void SysTimerISR(void) __attribute__ ((interrupt));
```

Also in **main.c**:

```
AIC_SMR12 = 0x00000027;
/* address of TC0 ISR to source vector reg. */
AIC_SVR12 = (unsigned long)SysTimerISR;
AIC_IECR = 0x00001000;       /* enable TC0 interrupt */
```

The PID (Peripheral ID) of the registers in the AIC (Advanced Interrupt Controller) that correspond to TC0 (PID12) is used to set up TC0 to use an interrupt. The address of the ISR gets copied to AIC_SVR12 (Source Vector Register 12). When TC0 triggers an interrupt, this address is loaded into the AIC_IVR which in turn gets loaded to the PC when our code jumps to the interrupt vector table.

The code then sets up TC0 in the same mode as our previous example with the exception that it now sets up the timeout value for one second, enables the interrupt on compare and starts the timer before entering the **while(1)** loop.

The **while(1)** loop sends out a global count to the serial port as fast as possible for display on a terminal emulator. The timer will cause the interrupt service routine to be run every second, when it does, it increments the global count.

In the ISR, the interrupt must be cleared by reading the TC0 status register (TC0_SR). At the end of the ISR, the AIC_EOICR (AIC End of Interrupt Command Register) must be written to in order to indicate to the AIC that the current interrupt is finished.

3.2.3 Nested Interrupt Code – Single Interrupt

The next example program, **PIT_Interrupt_Nested_Flash**, shows how nested interrupts are implemented in a C program. The assembly language startup file contains the code that will handle the interrupt. This code will re-enable interrupts, making it possible for the current interrupt to be interrupted by a higher priority interrupt. It will then call the interrupt service routine which is written in C and take care of returning properly to the main program when the ISR returns.

```
#include <stdio.h>
#include "at91sam7s.h"
#include "dbgu.h"

/* PIT ISR (interrupt service routine) */
void PIT_ISR(void);

unsigned int g_int_count = 0;

int main(void)
{
    char str[50];

    DBGUInit(DBGU_115200);

    DBGUTxMsg("\r\nTesting PIT with interrupt\r\n\r\n");
    /* setup PIT interrupt and ISR */
    AIC_SMR1 = 0x00000027;
    AIC_SVR1 = (unsigned long)PIT_ISR;
    AIC_IECR = 0x00000002;      /* enable PIT interrupt */

    PIT_MR = 0x030FFFFF;

    while (1) {
        sprintf(str, "count = %8d\r", g_int_count);
        DBGUTxMsg(str);
      }
}

void PIT_ISR(void)
{
    unsigned int pit_reg;

    g_int_count++;
    pit_reg = PIT_PIVR;         /* clear PIT interrupt */
}
```

This program enables the PIT interrupt and sets the PIT up to count to its maximum value. When the interrupt occurs, a global count is incremented. The value of the global count is continually sent to a terminal emulator via the DBGU serial port. Note that the PID of the PIT falls under PID1 which is the system PID and includes the PIT, RTT and other system devices.

The above example does not demonstrate nested interrupts, but only how to set up code to work with nested interrupts. The next example shows nested interrupts in operation.

3.2.4 Nested Interrupts

The **`PIT_TC0_Interrupts_Nested_Flash`** program is a combination of the previous two programs. The listing for this program is not included here so be sure to take a look at the source code.

In the program, the PIT and TC0 interrupts are enabled. The PIT ISR is set up as the highest priority interrupt and increments a global count that is sent to a terminal emulator by the main loop as our previous program did. The TC0 interrupt is set up as the lowest priority interrupt and is used to flash an LED on and off.

4. Digital Inputs

Digital inputs refer to the general purpose input pins of the microcontroller and are also known as ***discrete inputs***. We have seen the general purpose I/O pins on the microcontroller being configured as outputs, in this chapter we will configure them as inputs.

This chapter will show how to interface the following devices:

- Switches
- Optocouplers
- Keypads

4.1 Interfacing Switches

Push button switches, toggle switches, key switches, micro switches, reed switches and sensors such as proxy switches can all be interfaced to a microcontroller pin. If the switches are local, the interface is simpler. If the switches come from out in the field, extra circuitry is required in order to protect the microcontroller I/O pin.

4.1.1 Local Switches

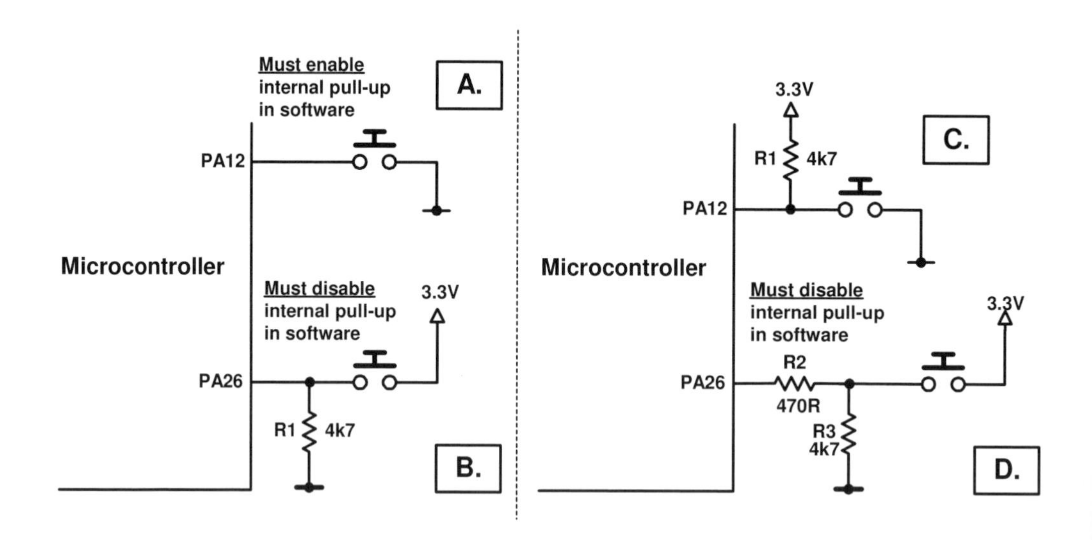

Figure 4.1: Interfacing a Switch Directly to the Microcontroller

The configurations shown in **figure 4.1** are best used for switches that are directly connected to the PCB that the microcontroller is attached to, or in the same box in which the microcontroller exists. Extra protection should be used when the switches are attached to the microcontroller from out in the field. This will be discussed shortly.

Figure 4.1 A. is the simplest method of connecting a switch to an AT91SAM7S microcontroller. The internal pull-up resistor must be enabled on the pin when using this configuration. In software, a logic 1 will be read when the switch is open and a logic 0 will be read when the switch is closed.

Figure 4.1 B. reverses the logic values that will be read on the switch. The internal pull-up resistor must be disabled when using this configuration. Pull-down resistor R1 is used to pull the pin low when the switch is open. A logic 0 is read in software when the switch is open and a logic 1 is read when the switch is closed.

Figure 4.1 C. works the same way as **figure 4.1 A.**, but uses a stronger external pull-up resistor than the internal pull-up resistor. In software, the internal pull-up resistor can be enabled or disabled.

Figure 4.1 D. is the same as **figure 4.1 B.**, but adds a protection resistor R2. This protection resistor can be used with any of the configurations in **figure 4.1**. The resistor protects the I/O pin from being shorted to the power or ground rails should the I/O pin accidentally be configured as an output and the switch attached to it closed.

4.1.2 Field Switches

Figure 4.2 shows some interface circuits that are more suited to switches that are located in the field, i.e. away from the microcontroller or the box that it is housed in.

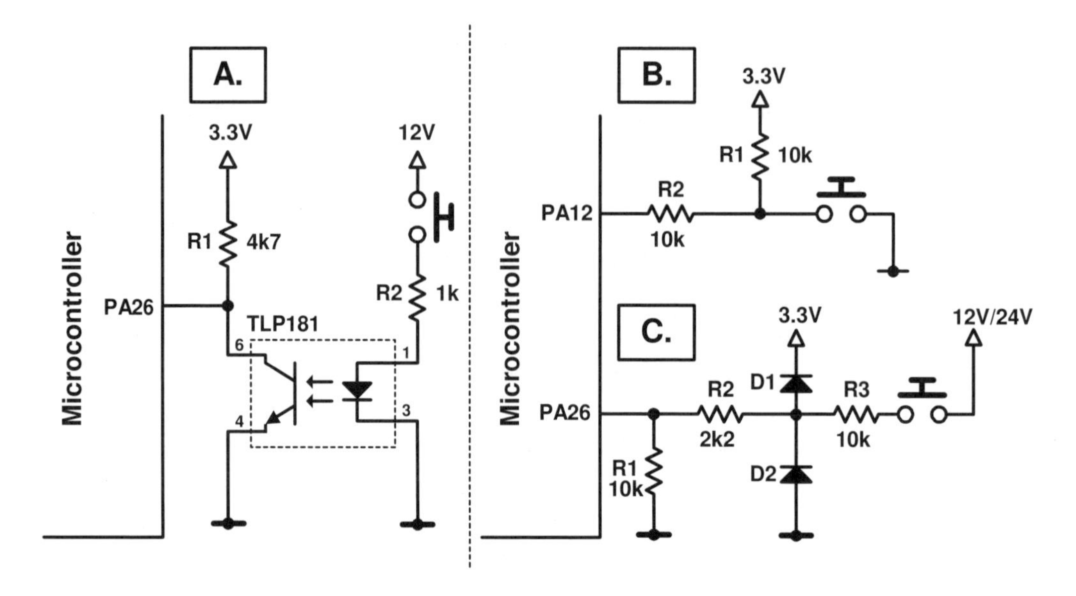

Figure 4.2: Switch Interface Circuits More Suited to Switches Connected in the Field

Figure 4.2 A. is the safest way to interface a switch to microcontroller – by using an optocoupler. Resistor R2 in the figure has been selected for use with 12V. This resistor must be changed according to the external voltage that is being switched. When the transistor in the optocoupler switches on, it will pull the microcontroller pin to within 0.2V of ground. This is adequate as the pin will register a logic 0 with up to 0.8V on the pin. The TLP181 used to test the circuit switched the pin to 0.11V.

Figure 4.2 B. can be used to interface a local or field switch to the microcontroller, although does not offer much protection to the microcontroller pin.

Figure 4.2 C. uses diodes to protect the input pin of the microcontroller from high voltages. When the switch is closed, current flows through R3 and D1 to the 3.3V supply. This drops most of the switched voltage across R3 and 0.7V across D1. The anode of D1 will now measure 4V relative to ground (3.3V + 0.7V). R1 and R2 act as a voltage divider that drops 3.3V of the 4V across R1, making it safe for the microcontroller pin. The 3.3V regulator must be able to sink the current flowing into it.

D2 is an optional diode that will conduct if a negative voltage is switched onto it. Most of the -0.7V across it will appear on the microcontroller pin. The microcontroller pin can handle an absolute maximum of only -0.3V on the AT91SAM7S, so the protection is not adequate but if replaced by a Schottky diode can be reduced to about -0.2V at low currents.

4.1.3 Software for Setting up and Reading Single Inputs

What the Software Must Do

On any microcontroller, the software must set the I/O pin up as an input and then read a register to determine if the pin is high or low so that the state of the switch will be known.

On the AT91SAM7S

The following program, **digital_input_int_RAM_dbgu** (or the Flash version of the program **digital_input_int_Flash_dbgu**), reads the state of the switch on port pin PA12 and switches the LED on and off on PA11 if the switch is closed or opened. The interrupt for the PIOA controller is enabled so that the number of times that the switch makes contact when closed or opened can be displayed on a terminal emulator program via the DBGU port. This demonstrates the “bounce” that occurs when a switch is closed or opened. Switch bounce occurs because a mechanical switch will not close or open cleanly, but will rather open and close a number of times very quickly when closed or opened – known as switch bounce.

```
#include <stdio.h>
#include "at91sam7s.h"
#include "dbgu.h"

#define PA11_OUT      (1 << 11)
#define PA12_IN       (1 << 12)

/* PIOA ISR (interrupt service routine) */
void PIOAISR(void);

unsigned int bounce_count = 0;
```

```
int main(void)
{
    char str[50];

    DBGUInit(DBGU_115200);

    /* setup PIOA interrupt and ISR */
    /* configure the Advanced Interrupt Controller (AIC) */
    /* edge triggered, highest priority interrupt */
    AIC_SMR2 = 0x00000027;
    /* address of PIOA ISR to source vector register */
    AIC_SVR2 = (unsigned long)PIOAISR;
    AIC_IECR = 0x00000004;         /* enable PIOA interrupt */

    /* enable the clock of PIOA */
    PMC_PCER = 0x00000004;
    /* configure PIO pin PA12 as input */
    PIOA_PER = PA12_IN;            /* enable uC pin as PIO   */
    PIOA_ODR = PA12_IN;            /* enable pin as input    */
    PIOA_PUDR = PA12_IN;           /* disable the pull-up resistor */
    PIOA_IER = PA12_IN;            /* enable PIOA interrupt */

    /* configure PIO pin PA11 as output */
    PIOA_PER  = PA11_OUT;          /* enable uC pin as PIO   */
    PIOA_PUDR = PA11_OUT;          /* disable the pull-up resistor */
    PIOA_OER  = PA11_OUT;          /* enable uC pin as output */

    while (1) {
        sprintf(str, "%d     \r", bounce_count);
        DBGUTxMsg(str);
        if (PIOA_PDSR & PA12_IN) {
            PIOA_CODR = PA11_OUT;
        }
        else {
            PIOA_SODR = PA11_OUT;
        }
      }
}

void PIOAISR(void)
{
    /* stores source of PIOA interrupt */
    volatile unsigned int pioa_status;
    /* had to read PIOA_ISR twice to reliably clear interrupt */
    pioa_status = PIOA_ISR;        /* clear the PIOA interrupt */
    pioa_status = PIOA_ISR;        /* clear the PIOA interrupt */
    bounce_count++;
}
```

In this program, the DBGU port is initialised first and then the input change interrupt is enabled. This interrupt will occur whenever there is a change in state on one of the input pins of the microcontroller i.e. from 0 to 1 or 1 to 0.

```
PMC_PCER = 0x00000004;
```

This line of code enables bit 2 (PID2) in the PMC_PCER register which enables the peripheral clock to the PIOA controller. The correct bit to set in the PMC_PCER register is found in the "Peripheral Identifiers" table in the AT91SAM7S datasheet. This table shows that the PIOA controller has a Peripheral ID (PID) of 2. (PMC = Power Management Controller, PCER = Peripheral Clock Enable Register). It is necessary to enable the clock to the PIOA controller in order to use the I/O pins as inputs.

```
PIOA_PER = PA12_IN;
```

Enables pin PA12 to be controlled by PIOA, same as for when setting up output pins.

```
PIOA_ODR = PA12_IN;
```

Setting the corresponding bit in the PIOA_ODR register enables the pin as an input. (ODR = Output Disable Register) Disabling the pin as an output enables it as an input.

```
PIOA_PUDR = PA12_IN;
```

Disables the internal pull-up on pin PA12, same as for the input pins. This may or may not be required, depending on the hardware configuration.

```
PIOA_IER = PA12_IN;
```

Enables the input change interrupt on PA12. (IER = Interrupt Enable Register)

The next lines in the program enable PA11 as an output as you have seen in the Digital Outputs chapter. In the **while(1)** loop, the global variable **bounce_count** is continually sent out of the DBGU serial port for display on a terminal emulator.

The `if` statement checks the state of the input pin PA12 and either sets or clears pin PA11 to switch the LED on or off. In the `if` statement the bit corresponding to pin PA12 is tested in the PIOA_PDSR to see if the switch attached to it is opened or closed (PDSR = Pin Data Status Register).

The code in the interrupt service routine, `PIOAISR()`, will run every time that the switch on PA12 is closed or opened, incrementing the global variable `bounce_count` each time. Because the switch contacts bounce, `bounce_count` will be incremented more that once each time the switch is closed or opened. The code in the `while(1)` loop is continually sending the count value to the terminal emulator for display.

Note that in the software programs dealing with switches, the bit value that is read for the switch state may be 1 or 0 depending on the configuration of the switch. Some configurations will cause a 1 to be read when the switch is open and a 0 when it is closed, others will work the opposite way around. When using the program to switch an LED, the LED may be wired in a current sourcing or current sinking configuration, which may also reverse the logic that is required to switch it on and off, so adjust the logic in each program according to your hardware configuration.

4.1.4 Switch Debounce Software

The procedure for debouncing a switch is as follows:

1) Detect a change in the switches state (from open to closed or closed to open).
2) Wait approximately 30ms for the switch contact bouncing to stop.
3) Read the state of the switch to see if it is closed or open. If the switch is closed, it is a valid closed state; if the switch is open it is a valid open state.

This procedure debounces the switch when it is closed and when it is opened as bouncing occurs during both of these operations. The `digital_input_RAM_debounce` program demonstrates how to debounce a switch in C code on the AT91SAM7S. The hardware used is as follows:

- A switch is attached to pin PA12 in the configuration shown in **figure 4.1 D**.
- An LED is attached to pin PA11 in current sinking configuration.
- Timer counter 0 (TC0) is used for the debounce time delay.

Only the code in **main()** is shown below:

```
int main(void)
{
  unsigned char switch_closed = 0;       } Variables used as flags
  unsigned char switch_changed = 0;

  PIOPinsEnable();
  TimerSetup();

  while (1) {
      /* check if switch state has changed */
      if ((PIOA_PDSR & PA12_IN) && !switch_closed) {
          /* switch changed from open to closed */
          switch_closed = 1;
          switch_changed = 1;      /* flag the change */
      }                                                      ① Switch
      else if (!(PIOA_PDSR & PA12_IN) && switch_closed) {   edge
          /* switch changed from closed to open */           detection
          switch_closed = 0;
          switch_changed = 1;      /* flag the change */
      }
      /* switch changed from closed to open or open to closed */
      if (switch_changed) {
          switch_changed = 0;
          TimerWait(30);          /* debounce timer */  } ② Debounce
          if (PIOA_PDSR & PA12_IN) {                        delay
              /* switch is debounced and closed */
              PIOA_CODR = PA11_OUT;
          }                                                  ③ Read valid
          else {                                             "debounced"
              /* switch is debounced and opened */           switch state
              PIOA_SODR = PA11_OUT;
          }
      }
  }
}
```

The program uses the **switch_closed** variable in the first **if - else if** construct to detect if the switch has changed state i.e. has been closed from the open state or opened from the closed state. This is also known as edge detection as the transition between states is being detected. If a change has occurred, this is flagged by setting the **switch_changed** variable.

The second `if` statement will only run if a change in the switch state was detected. A delay of 30ms will take place if the switch was closed or opened in order to wait for the switch contacts to stop bouncing. The state of the switch is now read again in order to determine if the switch is closed or open. At this stage the switch has been debounced.

4.2 Interfacing Keypads

A keypad can be easily interfaced to the AT91SAM7S microcontroller as shown in **figure 4.3**.

Circuit Description

Diodes **D1 to D4** in the **figure 4.3** protect the outputs pins from being shorted to each other should two or more keys in a column be pressed at the same time. The diodes used can be any general purpose diodes such as the **1N4148**. Resistors **R1 to R3** are pull-up resistors for the keypad columns.

Principle of Operation

To use the keypad, a software program is written to drive one of the output pins (PA0 to PA3) low at a time and the rest of the output pins high. For example we will start by driving PA0 low. Now if key 1, 2 or 3 is pressed in, we will read a 0 on pins PA4, PA5 or PA6 respectively. In the software program we know that we are driving pin PA0 low and that it is connected to keys 1, 2 and 3, so we will know which key is pressed when we read PA4 to PA6.

The next row is scanned in a similar way – PA1 is now driven low, PA0, PA2 and PA3 are driven high. PA4 to PA6 are now read to see if key 4, 5 or 6 has been pressed, and so on for the remaining two columns.

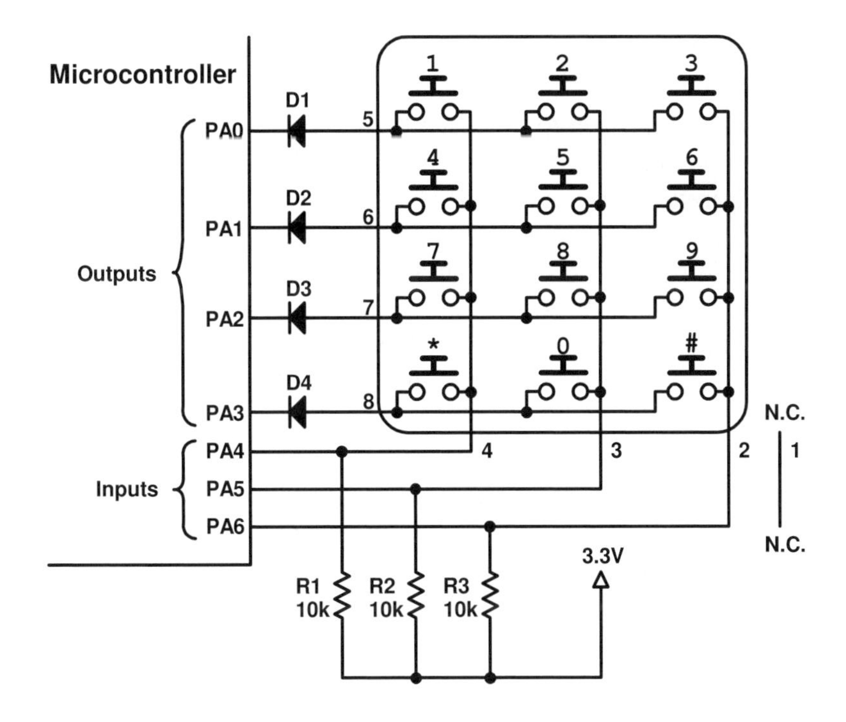

Figure 4.3: A 4 Row by 3 Column Keypad Interfaced to the AT91SAM7S

Protection Diodes

When for example PA0 is driven low and PA1 is driven high and keys 1 and 4 are pressed simultaneously, we would be shorting an output pin that is driven high to an output pin that is driven low. The diodes protect the output pins from being shorted in this way. A disadvantage of the diodes is that when a key is pressed with an output pin connected to it that is driving low, the corresponding input pin to which the key connects it will read this low voltage as 0.7V instead of 0V because of the diode. This is still within the 0.8V that the input pin requires in order to register a low, but is uncomfortably close. The diodes could be replaced with **470Ω** resistors and still offer some protection while bringing the input pins closer to 0V when connected to an output pin that is driven low. Schottky diodes with their low forward voltage could also be used.

4.2.1 Keypad Software – Consecutive I/O Pin Connections

The microcontroller pins used to interface the keypad to the microcontroller in **figure 4.3** are conveniently situated next to each other and all at the bottom of the word that is read or written when the I/O pins are accessed. This makes the software easy to write. We will first look at a program that uses the pin configuration in the figure and then look at an example of what to do when the pins connected to the keypad are scattered around in the I/O word of the microcontroller. This may result because there may be internal peripheral devices on some or all of the pins that are needed in an embedded system and so cannot be assigned to the keypad.

The **`digital_input_keypad12_RAM`** program operates the 12 key keypad of **figure 4.3**. The program does not use any interrupts and does not debounce the keys of the keypad. It sends the character of the key pressed out of the DBGU RS-232 serial port for capture on a terminal emulator program attached to the PC. The character will be sent continuously as long as the key is held in.

```
#include "at91sam7s.h"
#include "dbgu.h"

#define KEY_OUT        0x0000000F    Output and Input pins connected
#define KEY_IN         0x00000070    to keypad defined

void KeypadInit(void);                                   Keypad Driver
char KeypadGetKey(void);                                 Function Prototypes
unsigned char KeypadGetColumn(unsigned char code);

const char g_keypad_lookup[4][3] =  {{'3', '2', '1'},
                                     {'6', '5', '4'},
          Store table in read-       {'9', '8', '7'},
          only memory                {'#', '0', '*'}};
                                                 Global lookup table
int main(void)                                   used by keypad driver
{                                                function
    char keypad_key = 0;
    char out_msg[] = "k\r\n";

    DBGUInit(DBGU_115200);
    KeypadInit();
```

```
    DBGUTxMsg("\r\n*** Keypad Test Program ***\r\n\r\n");
    while (1) {
        keypad_key = KeypadGetKey();
        if (keypad_key != 0) {
            out_msg[0] = keypad_key;
            DBGUTxMsg(out_msg);
        }
    }
}
```

How the `digital_input_keypad12_RAM` program works

The program uses a string, **`out_msg[]`**, to display the keypad key that is pressed. At the top of the program, **`out_msg[]`** is initialised with the string "k\r\n". The 'k' is just a place holder as the character of the key pressed on the keypad will be substituted for this character when a key is pressed. This was done so that **`sprintf`** or a similar function would not be needed. The "\r\n" simply moves the cursor to the beginning of the next line on the terminal emulator program.

At the beginning of the program, the DBGU port and keypad hardware are initialised, a message is sent to the terminal emulator and the **`while(1)`** loop is entered.

In the **`while(1)`** loop, the **`KeypadGetKey()`** function is called. This function will return 0 if no key was pressed on the keypad, or else the ASCII code for the key that was pressed.

The **`if`** statement will only run if a key was pressed i.e. a non-zero value was returned from **`KeypadGetKey()`**. Inside the **`if`** statement, the ASCII key value returned is substituted into the first element of the **`out_msg[]`** string and then the modified string is sent out of the DBGU serial port to be captured on a terminal emulator.

How the keypad driver functions work:
The keypad driver consists of three functions: **`KeypadInit()`**, **`KeypadGetKey()`** and **`KeypadGetColumn()`**.

The **KeypadInit()** function initialises the input and output pins used to interface the keypad:

```
void KeypadInit(void)
{
    /* enable the clock of PIOA (Parallel I/O controller A) */
    PMC_PCER = 0x00000004;

    /* configure input and output pins for the keypad */
    PIOA_PER  = (KEY_OUT | KEY_IN);
    PIOA_ODR  = KEY_IN;
    PIOA_OER  = KEY_OUT;
    PIOA_OWER = KEY_OUT;
}
```

Firstly the clock to the PIO controller is enabled in order to use the input pins. Pins PA0 to PA6 are configured as PIO pins, pins PA4 to PA6 are configured as inputs and PA0 to PA4 are defined as outputs. This should all be familiar to you from previous examples. What is new in this program is the initialisation of the PIOA_OWER (OWER = Output Write Enable Register). The bits that are set in this register (the bits corresponding to pins PA0 to PA3 are set in the program) can all be written to simultaneously by writing to the PIOA_ODSR (ODSR = Output Data Status Register). Previously we set and cleared individual bits, but now by writing to the PIOA_ODSR we can set and clear bits simultaneously with a single write to this register. The only bits that will be affected by a write to this register are the ones that were enabled by writing to the PIOA_OWER.

The **KeypadGetKey()** function returns the ASCII character of the key pressed or 0 if no key was pressed:

```
char KeypadGetKey(void)
{
    unsigned int  out_mask = 0x00000001;
    unsigned char in_code = 0;
    unsigned char key_col = 0;
    int row_num;
    char key_val = 0;
```

```
    for (row_num = 0; row_num < 4; row_num++) {
        PIOA_ODSR = ~out_mask;
        in_code = ~(unsigned char)((PIOA_PDSR >> 4) | 0xF8);
        if (in_code) {
            key_col = KeypadGetColumn(in_code);
            key_val = g_keypad_lookup[row_num][key_col];
            break;
        }
        out_mask <<= 1;
    }

    return key_val;
}
```

The **for** loop in this function writes a 0 to each row pin on the keypad in turn, starting with pin PA0. All other pins remain high. I.e. on the first pass through the loop PA0 is low, the other three pins are high, on the second pass through the loop, PA1 is low, all other pins are high, etc. It does this by using the **out_mask** variable which is set to an initial value of 1. At the bottom of the loop the value in this variable is left shifted by one. In other words it is moving the 1 from PA0 to PA1, to PA2 and finally to PA3. When this variable is written out to PA0 to PA3, it is inverted using the bitwise complement operator (~) so that a 0 is sent to the desired pin and a 1 to all other pins. This mask value is written to PA0 to PA3 simultaneously by writing to the PIOA_ODSR which was set up for this in the keypad initialisation function.

After clearing a single row pin of the keypad and setting all the rest, the column pins are all read simultaneously by reading from the PIOA_PDSR (PDSR = Pin Data Status Register). The problem with reading the row pins is that they occupy bits 4 to 6 of the PIOA_PDSR. To solve this problem we right shift them by 4 bits:

```
in_code = ~(unsigned char)((PIOA_PDSR >> 4) | 0xF8);
```

This line right shifts the read value by 4, then sets all remaining bits in the byte that the read value occupies by using the OR operator. The value is then cast to 8 bits and inverted with the bitwise complement operator. The top 24 bits of this value are truncated by casting it to and copying it to an 8 bit variable – **in_code**.

This process is done because firstly the value read back from the columns is offset and secondly it is inverted i.e. the column that is activated will be read as a 0. For the four values that could be read back, it changes them as follows (value read (binary) – converted value (binary)):

```
0x000000F0 (1111 0000) - 0x00 (0000 0000) No key pressed
0x000000E0 (1110 0000) - 0x01 (0000 0001) Column PA4 key pressed
0x000000D0 (1101 0000) - 0x02 (0000 0010) Column PA5 key pressed
0x000000B0 (1011 0000) - 0x04 (0000 0100) Column PA6 key pressed
```

This converted value is checked in the **if** statement to see if it is non-zero. If it is non-zero, it means that a key was pressed, if it is zero, a key was not pressed and nothing further needs to be done – the next row can be activated and checked.

If a key was pressed, the helper function **KeypadGetColumn()** is called in order to convert the number stored in the variable **in_code** into a value between 0 and 2 for use in the lookup table.

We now know which row we are busy with because it is stored in the **for** loop variable **row_num** and we know which column was activated by the pressed key as it was returned by **KeypadGetColumn()** and stored in the variable **key_col**. The lookup table **g_keypad_lookup[4][3]** is a global two dimensional array created to look up the ASCII equivalent of the key that was pressed. The table consists of 4 arrays of 3 characters each.

The 4 arrays represent the 4 rows and the 3 characters represent the 3 columns. Because we know both the row and column that is activated, we can get the correct value from the lookup table, break out of the **for** loop and return it:

```
key_val = g_keypad_lookup[row_num][key_col];
```

This program was relatively simple and did not include any debouncing or interrupts as it was written to demonstrate the principle of operating a keypad.

4.2.2 Keypad Software – Non-consecutive I/O Pin Connections

Figure 4.4 shows the same keypad as **figure 4.3**, but this time it is connected to different pins that are spread around the microcontroller. This may occur because some of the internal hardware devices attached to some of the microcontroller pins may be needed in an application, making it impossible to get a neat sequence of consecutive pins to interface to the keypad. The software drivers for the keypad will have to change because of this, but the driver functions still perform the same task in the main program.

A copy of the **digital_input_keypad12_RAM** project was made and the name changed to **digital_input_keypad12_2_RAM**. The following changes were made to the new project:

At the top of **main.c**, bit definitions were made for the new pins connected to the keypad:

```
#define PA4_OUT      (1 << 4)
#define PA7_OUT      (1 << 7)
#define PA8_OUT      (1 << 8)
#define PA11_OUT     (1 << 11)
#define PA12_IN      (1 << 12)
#define PA26_IN      (1 << 26)
#define PA30_IN      (1 << 30)

#define KEY_OUT      (PA4_OUT | PA7_OUT | PA8_OUT | PA11_OUT)
#define KEY_IN       (PA12_IN | PA26_IN | PA30_IN)
```

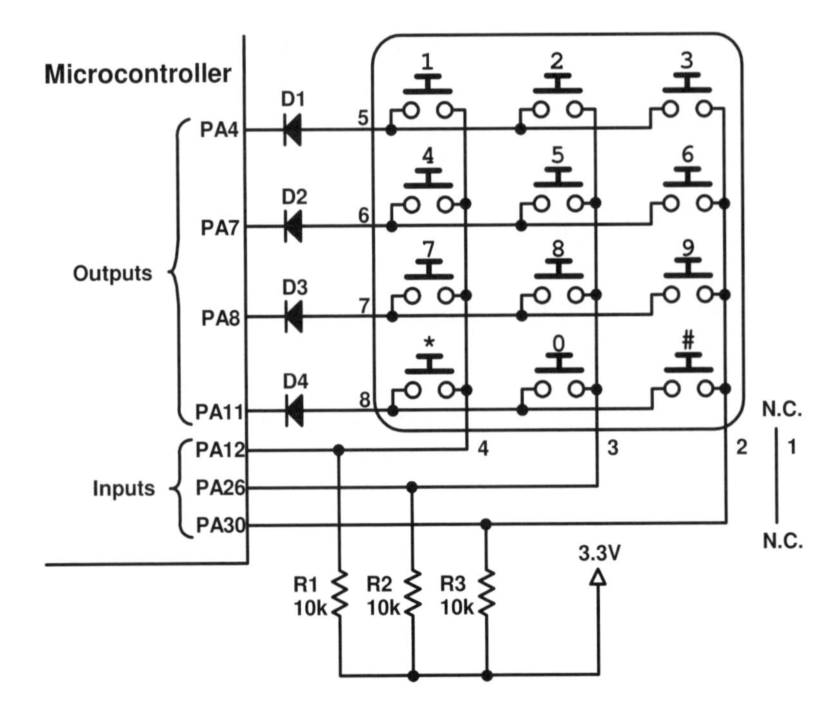

Figure 4.4: Same Keypad as Figure 4.3, but Connected to Different I/O Pins

It was easier to use these bit definitions than to work out a single hexadecimal number to initialise the input and output pins. The bit definitions were also used in the keypad driver functions.

It was necessary to change the order of the ASCII numbers in the keypad lookup table – this was actually done last after first testing the modified keypad drivers:

```
const char g_keypad_lookup[4][3] =  {{'1', '2', '3'},
                                     {'4', '5', '6'},
                                     {'7', '8', '9'},
                                     {'*', '0', '#'}};
```

The **KeypadInit()** function remained unchanged because of the **KEY_OUT** and **KEY_IN** definitions. The **KeypadGetColumn()** function remained unchanged.

The following changes were made to the **KeypadGetKey()** function:
The **out_mask** variable was changed into a lookup table:

```
unsigned int  out_mask[] = {PA4_OUT, PA7_OUT, PA8_OUT, PA11_OUT};
```

This was done because the output pins connected to the keypad are not sequential, so the mask could not be shifted by one to switch a single pin each time. The look up table hard codes the pins in the sequence that they will be switched. It is not necessary to shift the mask now as the **row_num** variable is used as an index into the lookup table.

An additional variable, **in_code_temp**, was required in order to capture the full 32-bit word read from the PIOA_PDSR containing the values of the 3 column pins. The variable ensures that all the input pin states are captured as they are distributed throughout this 32-bit register.

Inside the **for** loop, the following lines changed:

```
PIOA_ODSR = ~out_mask[row_num];
```

The look up table **out_mask** is used instead of the variable **out_mask**.

```
in_code_temp = PIOA_PDSR;
in_code = (unsigned char)(((in_code_temp & PA12_IN) >> 12) |
                          ((in_code_temp & PA26_IN) >> 25) |
                          ((in_code_temp & PA30_IN) >> 28));
in_code = ~(in_code | 0xF8);
```

The entire PIOA_PDSR register is captured to **in_code_temp** and then the bits representing the columns of the keypad are shifted to the bottom 3 bits of the word and stored in **in_code**. The data in **in_code** is then changed to the correct format for checking if a key was pressed and for use with the **KeypadGetColumn()** function as was done in the original program.

4.2.3 Keypad Software with Switch Debouncing

The **digital_input_keypad12_2_RAM** program was modified to create a new program that adds debouncing of the keypad switches. The new program is called **digital_input_keypad12_2_RAM_debounce**, only the **while(1)** loop of the program is shown here:

```
while (1) {
    keypad_key = KeypadGetKey();
    if (keypad_key != 0) {
        out_msg[0] = keypad_key;
        DBGUTxMsg(out_msg);
        TimerWait(30);  /* debounce delay for key pressed */
        /* wait for key to be released */
        while (KeypadGetKey() != 0);
        TimerWait(30);  /* debounce delay for key released */
    }
}
```

The timer functions **TimerSetup()** and **TimerWait()** that use timer counter TC0 were added to the program. **TimerSetup()** sets up and starts the timer at the beginning of the program.

The **if** statement of the **while(1)** loop will only be entered if a key on the keypad is pressed. As soon as a key is pressed, the key is sent out of the DBGU serial port to the terminal emulator for display. A 30ms delay now occurs to debounce the keypad switch. Note that if the switch bounced at the top of the **while(1)** loop when the keypad is being scanned by **KeypadGetKey()**, this function would simply return 0 and then be re-run as it is inside a continuous loop. Any bounce would therefore be skipped.

After the debounce delay, the program scans the keypad continuously until the key that was pressed is released. When the key is released, a second 30ms delay occurs in order to debounce the key release.

4.2.4 The Concept of Drivers

Keypad drivers were used to initialise the I/O pins used to interface to the keypad as well as to get the key that was pressed from the keypad. By dividing the software into drivers, the software not only becomes easier to read and understand, but is more portable. Portable means that the program can be moved easily to a different hardware platform. For example, if one of the keypad programs was to be moved to a different manufacturers microcontroller, only the driver code would need to be changed in order to interface to the different hardware.

In the keypad example programs of this chapter, the driver functions were included in the `main.c` file. This was to make the program simpler to understand. It is better programming practice to move the keypad driver functions to their own C source file. They can then be easily re-used by copying the keypad driver file to a new project.

Drivers are often represented diagrammatically as a layer as shown in **figure 4.5**.

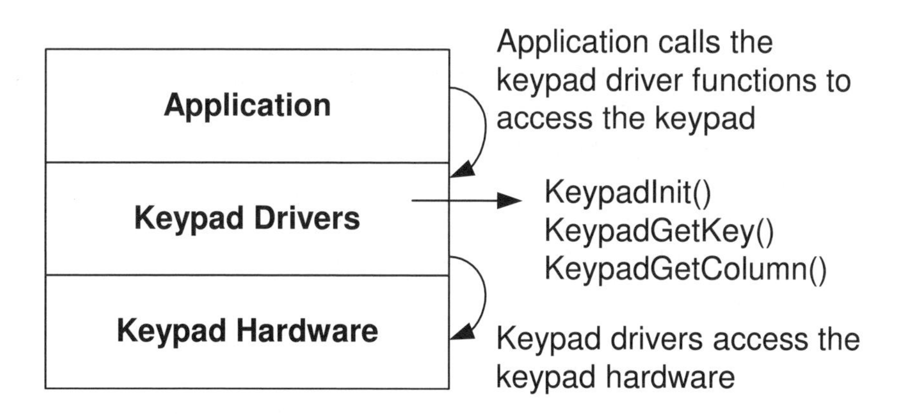

Figure 4.5: Drivers Shown Diagrammatically

The figure shows that the application layer accesses the hardware layer through the keypad driver layer. In practice this is done by calling the keypad driver functions from the main program.

5. LCD and Seven Segment Displays

The most common LCD displays are LCD character displays that are capable of displaying alphanumeric data as well as some special symbols.

Seven segment displays are LED displays used for displaying numbers such as quantity or time.

5.1 LCD Character Displays

Character LCD displays based on the **Hitachi HD44780 LCD controller and driver IC** are very common and available from various manufacturers.

Character LCD displays are available with different numbers of rows and characters in each row. E.g. a 16 × 2 LCD display has two rows that can display 16 characters each, a 20 × 4 LCD display has 4 rows each able to display 20 characters.

These displays are available with or without backlighting and operate from a 5V power supply. It is possible to drive these displays from a 3.3V microcontroller as a logic high applied to one of the logic pins of the LCD is valid at a voltage of 2.2V, although a 5V power supply is still required to power the LCD. Backlighting refers to lighting behind the display that makes the display easier to read in poor lighting conditions. Some displays have backlighting built in, others have no backlighting.

The displays can operate in either 4 bit or 8 bit mode. 4 bit mode uses only 4 data lines, while 8 bit mode uses 8 data lines. The 4 bit interface operates slower than 8 bit, but frees up 4 microcontroller pins. Some control lines are also necessary to operate the LCD. 3 control lines are present on the display, but if we do not want to read anything back from the LCD, we can use only 2 of these lines. An LCD operating in 8 bit mode will require a total of 10 microcontroller pins and 6 microcontroller pins will be required in 4 bit mode.

5.1.1 LCD Interfacing Hardware

Figure 5.1 shows how to connect a character LCD display to an AT91SAM7S microcontroller. **Figure 5.2** shows the connections on a 2 × 16 character LCD display.

Some examples of 2 × 16 character LCD displays with backlighting are:

MSC-C162DYLY-4N from TRULY
PC1602LRS-FWA-B from POWERTIP

These two displays are both mechanically and electrically compatible – they are drop-in replacements for each other. The Truly and Powertip displays are physically pin for pin compatible, but the pin numbers are labelled differently. The left pin column in **figure 5.2** contains the Truly pin numbers and the right pin column contains the Powertip pin numbers. **Figure 5.2** explains the function of each pin of the LCD.

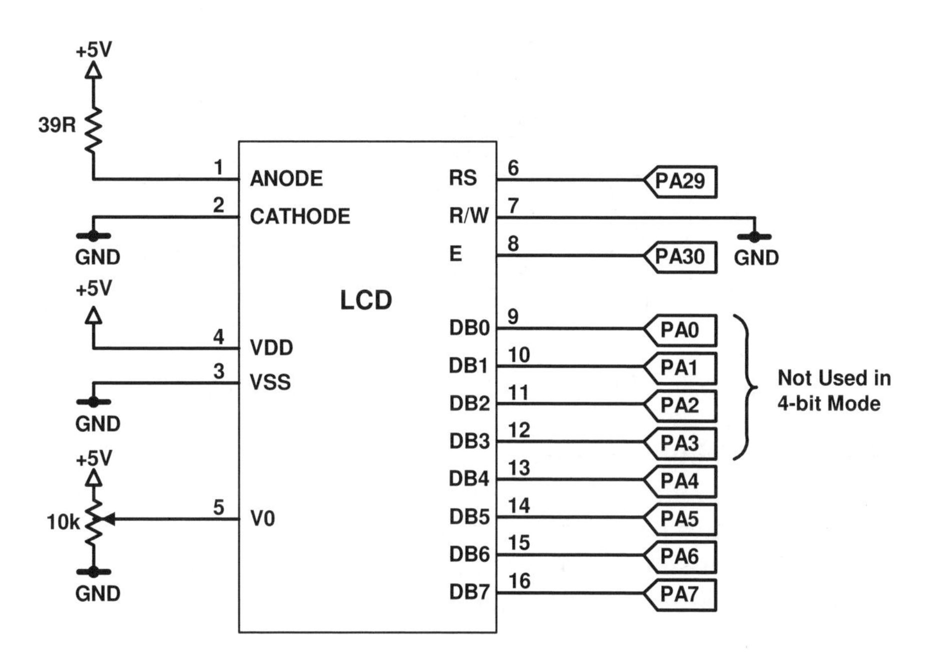

Figure 5.1: 16 × 2 Character LCD Display Connections to Microcontroller

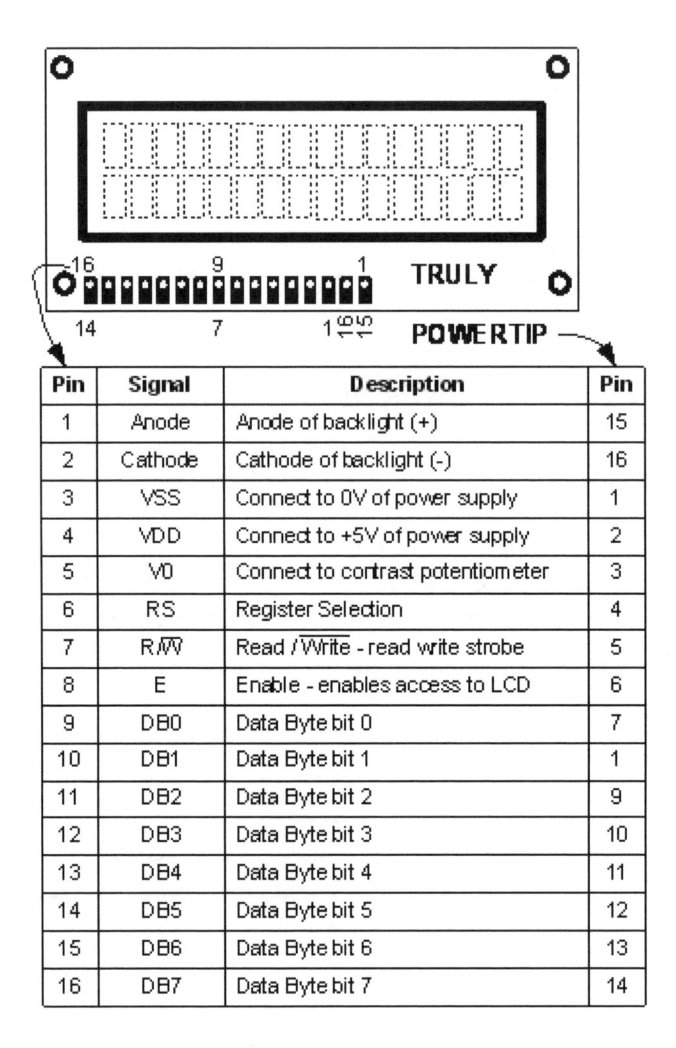

Pin	Signal	Description	Pin
1	Anode	Anode of backlight (+)	15
2	Cathode	Cathode of backlight (-)	16
3	VSS	Connect to 0V of power supply	1
4	VDD	Connect to +5V of power supply	2
5	V0	Connect to contrast potentiometer	3
6	RS	Register Selection	4
7	R/$\overline{W}$	Read / $\overline{\text{Write}}$ - read write strobe	5
8	E	Enable - enables access to LCD	6
9	DB0	Data Byte bit 0	7
10	DB1	Data Byte bit 1	1
11	DB2	Data Byte bit 2	9
12	DB3	Data Byte bit 3	10
13	DB4	Data Byte bit 4	11
14	DB5	Data Byte bit 5	12
15	DB6	Data Byte bit 6	13
16	DB7	Data Byte bit 7	14

Figure 5.2: 16 × 2 Character LCD Display Connections

5.1.2 LCD Interfacing Software

Character LCD displays need to be initialised after power-up by writing various instructions to them as shown in **figure 5.3**. The instructions used for initialisation and operating the LCD are shown in **figure 5.4**.

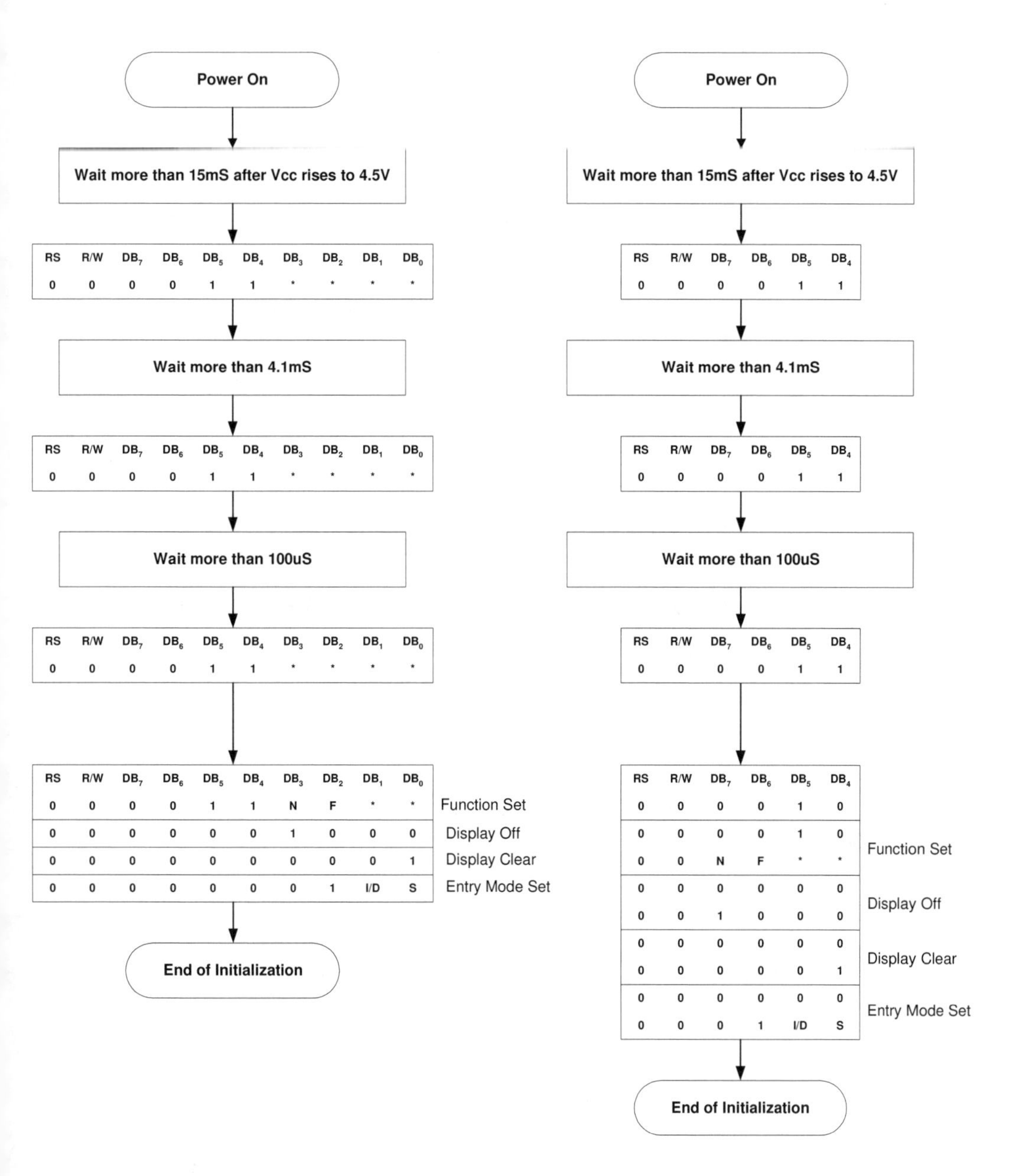

Figure 5.3: LCD Initialisation Instructions for 8-bit (Left) and 4-bit (right) Modes of Operation

Instruction	Code										Description	Executed Time (max.) fosc = 250kHz
	RS	R/W	DB7	DB6	DB5	DB4	DB3	DB2	DB1	DB0		
Clear Display	0	0	0	0	0	0	0	0	0	1	Clears entire display and sets DD RAM address 0 in address counter.	1.64ms
Cursor to Home	0	0	0	0	0	0	0	0	1	*	Returns the cursor to the home position (address 0). Retruns display being shifted to original position.	1.64ms
Entry Mode Set	0	0	0	0	0	0	0	1	I/D	S	Sets cursor move direction and specifies shift of display. I/D = 1: increment; ID = 0: decrement S = accompanies display shift	40us
Display On/Off	0	0	0	0	0	0	1	D	C	B	D - sets ON/OFF of entire display C - cusor ON/OFF B - blink of cursor poition character	40us
Cursor/Display Shift	0	0	0	0	0	1	S/C	R/L	*	*	S/C = 1: Display shift; S/C = 0: Cursor move R/L = 1: Shift to right; R/L = 0: Shift to Left	40us
Function Set	0	0	0	0	1	DL	N	F	*	*	DL (sets interface data length) - DL = 1: 8 bits; DL = 0: 4 bits N (number of display lines) - N = 1: 2 lines; N = 0: 1 line F (character font) - F = 1: 5 x 10 dots, F = 0: 5 x 7 dots	40us
Set CG RAM Address	0	0	0	1	ACG						Sets CG (character generator) RAM address. CG RAM data is sent and received after this setting.	40us
Set DD RAM Address	0	0	1	ADD							Sets DD (display data) RAM address. DD RAM data is sent and received after this setting.	40us
Busy Flag & Address	0	1	BF	AC							Reads Busy Flag (BF) indicating internal operation is being performed and reads address counter (AC) contents. BF = 1: internally operating; BF = 0: can accept instruciton	0us
Write to CG or DD RAM	1	0	Write Data								Writes data into DD RAM or CG RAM.	40us
Read Data from CG or DD RAM	1	1	Read Data								Reads data from DD RAM or CG RAM.	40us

Figure 5.4: Character LCD Display Instructions

After initialisation, the LCD is ready for operation and characters can be sent to the LCD for display. With the hardware configuration of **figure 5.1**, it is not possible to read data back from the display because the R/W input of the LCD is permanently set low (write only). The busy flag would normally be read to see if the LCD can accept new data or instructions, but we can wait for a time period between sending data or instructions instead.

To write an instruction to the LCD display, such as clear display, cursor on, etc. the instruction is written on DB0 to DB7 with the RS pin of the LCD pulled low. To write data that will be displayed on the LCD, the data is written on DB0 to DB7 with the RS pin pulled high. In 4-bit mode, the 8-bit instruction or data is written in two halves to DB4 to DB7.

Writing data or an instruction to the LCD
In **8-bit mode**, the procedure for writing data or an instruction to the LCD is as follows:

1. Set the RS line if writing data, or clear the RS line if writing an instruction
2. Set the E line
3. Place the data or instruction on DB7 to DB0
4. Clear the E line

In **4-bit mode**, the procedure for writing data or an instruction to the LCD is as follows:

1. Set the RS line if writing data, or clear the RS line if writing an instruction
2. Set the E line
3. Place the upper nibble (D7 to D4) of data or instruction on DB7 to DB4
4. Clear the E line
5. Set the E line
6. Place the lower nibble (D3 to D0) of data or instruction on DB7 to DB4
7. Clear the E line

Writing characters to the LCD display
The LCD display has internal data display ram (DD RAM). When a character is written to the DDRAM, it is displayed on the LCD. On the 2 × 16 LCD display, the first line of the display corresponds to the DD RAM addresses 0 to 15 (0x00 to 0x0F). The second line of the display corresponds to 64 to 79 (0x40 to 0x4F). In other words, to write a character to the first position on the second line of the display, we set the DD RAM address to 0x40 and then send the data byte.

The address is then automatically incremented, so that the next character written to the LCD will appear to the right of the first. To set the DD RAM address, an instruction is written to the LCD (RS = 0) containing the desired address. The instruction can be found in the row called **Set DD RAM Address** of **figure 5.4**. This instruction must have DB7 = 1 and DB6 to DB0 contain the address to set the DDRAM to.

LCD example programs

Two LCD example programs have been provided:

1) The first program, `LCD_Char_2x16_8bit_RAM`, is written for the configuration of **figure 5.1** – 8 bit mode, interfaced to consecutive microcontroller pins for the data bus.
2) The second program, `LCD_Char_2x16_4bit_RAM`, is the same as the first, but initialises the LCD in 4 bit mode. In **figure 5.1**, microcontroller pins PA0 to PA3 are disconnected from the LCD display for this example – LCD connections DB0 to DB3 are not used.

The two programs do the same thing. They initialise the LCD display, the first in 8 bit mode and the second in 4 bit mode. The programs then display a menu on a terminal emulator connected to the DBGU port. Numbers 1 to 6 can be sent from the terminal emulator to operate various LCD functions, such as switching the cursor on and off, displaying a message, etc.

The only differences between the two programs are in the `lcd.c` file. The `LCDInit()` functions differ as shown in **figure 5.3**. The 4 bit mode program adds the function `LCDWriteNibble()` and uses it in the LCD initialisation as the first few writes to the LCD are nibble writes. The second difference is in the `LCDWrite()` function: in the 4 bit mode program, two nibble writes are made in this function to send one byte.
The LCD functions of the two programs contained in `LCD.c` are as follows:

<u>**Low level functions:**</u>

These four functions are the low level drivers of the LCD and are called by the user functions to operate the LCD.

```
static void LCDWrite(unsigned char RS, unsigned char data)
```
The **RS** parameter of this function corresponds to the RS line of the LCD and is set to 0 to write an instruction to the LCD or to 1 to write data for display on the LCD. The labels **INSTR_WR** and **DATA_WR** are defined at the top of the **lcd.c** file to be passed to the **RS** parameter and make it easier for programmers to read.

The **data** parameter is passed the instruction from **figure 5.4** if an instruction is to be sent or the character to display if data is to be written.

```
static void LCDWriteNibble(unsigned char RS, unsigned char data)
```
This function is only present and used in the 4 bit mode program. It does exactly the same as the previous function, but sends only a nibble.

```
static void LCDDelay(unsigned int count)
```
Millisecond delay function used by other LCD functions.

```
static void uSecDelay(void)
```
Microsecond delay function used by other LCD functions.

An improvement to the **lcd.c** functions would be to use an internal timer to generate the timing delays required by the LCD. The above two timing functions use up time by counting in a loop. This is not accurate and will change (run slower) if the program is run from Flash memory.

User functions:

```
void LCDInit(void)
```
Initialises the LCD according to the instructions in **figure 5.3**.

```
void LCDPut(unsigned char x, unsigned char y, char *string)
```
Writes the string to the LCD display starting at the x/y position given. x is the horizontal position of the first character of the string starting at 0, y is the vertical position of the string on the LCD, starting at 0 i.e. 0 for y will put the string on line 1 of the LCD, 1 will put the string on line 2 of the LCD.

```
void LCDClear(void)
```
Clears the LCD display and moves the cursor to the home position.

```
void LCDDisplayOption(unsigned char option)
```
Controls the on/off of the display and the cursor.

```
void LCDgotoxy(unsigned char x, unsigned char y)
```

Positions the cursor at the x/y position passed to it.

5.2 Seven Segment Displays

Seven segment displays are so called as they are made up of 7 LEDs from which the numbers 0 to 9 can be displayed. Most seven segment displays also have a decimal point after the digit as shown in **figure 5.5**.

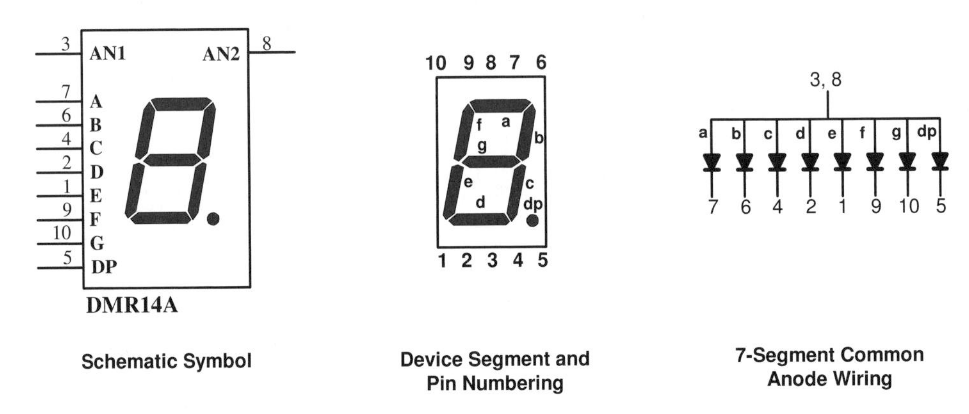

Figure 5.5: Example of a Seven Segment Common Anode Display from SunLED

Note that not all seven segment displays have the same pin-out as the one shown in the figure. Common anode and common cathode displays are available. Common anode displays have all of the anodes of the display's LEDs joined together, common cathode displays have all of the cathodes joined together. **Figure 5.6** shows two ways of interfacing a single digit common anode seven segment display to a 3.3V microcontroller. The same method could be used to interface to a 5V microcontroller.

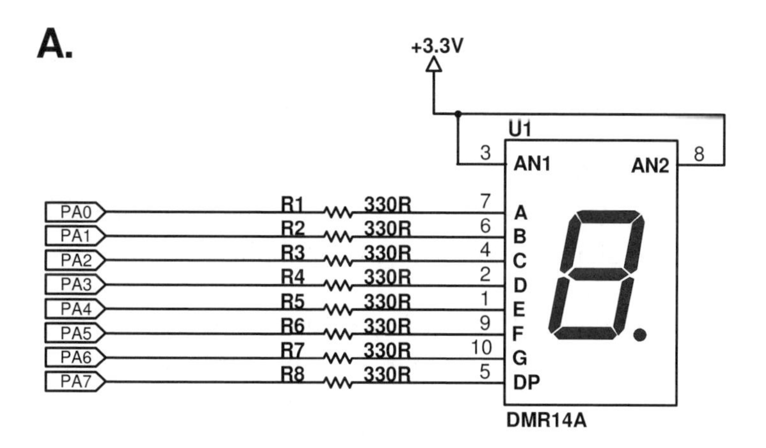

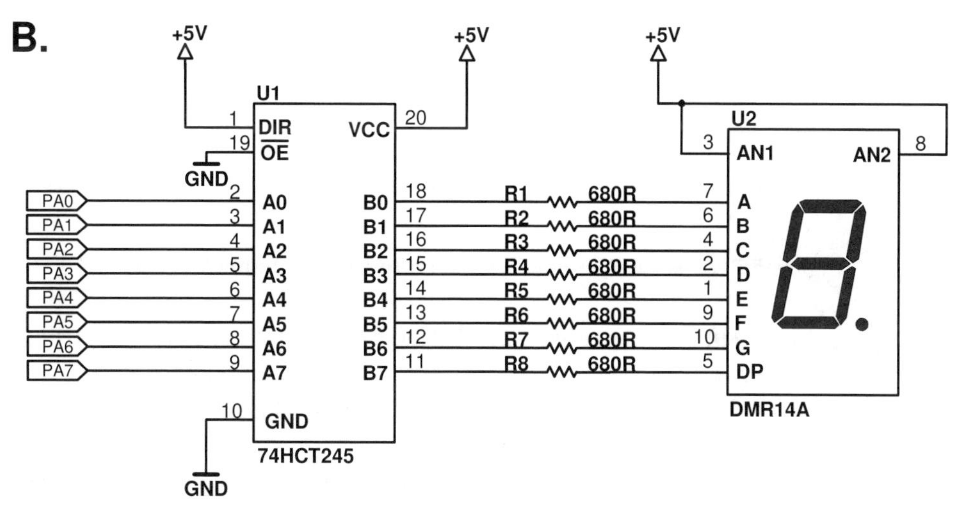

Figure 5.6: Driving a Single 7-Segment Digit (Common Anode) A) Directly from Microcontroller, B) From 5V Through a Buffer IC

Figure 5.6 A. shows that the display can be interfaced directly to the microcontroller pins. This is the same as interfacing 8 LEDs to the microcontroller and the calculations on LED interfacing given in **chapter 2** apply.

Figure 5.6 B. is an example of how to drive a 5V device with a 3.3V microcontroller. The 74HCT logic family are 5V devices, but when driven will register a logic high at 2V. In the figure, the 74HCT245 octal bus transceiver is acting as a buffer as it is forced to buffer in one direction only. Because 74245s are bi-directional buffers, it may be more worth while to stock this part than an octal buffer as it can be forced to buffer either left to right or right to left if desired. The 74HCT245 can deliver 6mA of current per pin.

A 74**HC**245 can be powered by a voltage from 2V to 6V. It could be used in a 3.3V circuit and powered by 3.3V, but if powered from 5V could not be used to drive a 5V load from a 3.3V microcontroller. The reason for this is that the 74HC logic family will only register a logic high input at 0.7 × Vcc, which is 3.5V when a Vcc of 5V is used.

Figure 5.7 shows which LED segments to switch on to display the numbers 0 to 9.

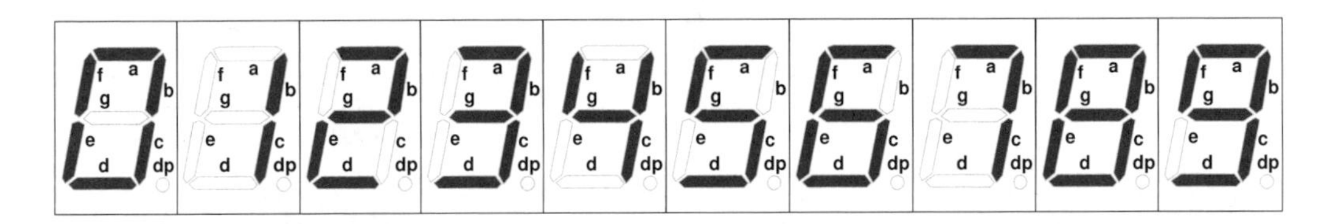

Figure 5.7: LED Segments to Switch on to Display Numbers 0 to 9

For the circuits of **figure 5.6**, a table can be made that displays what to write to the microcontroller port to display each number on the seven-segment display. This information is shown in **table 5.1**. Note that this will only work if the seven-segment display is wired to the port as shown with segment **a** wired to port pin **PA0**, segment **b** wired to port pin **PA1**, etc. If you wire the display to different port pins, you will have to change the value that gets written to the port to be able to display the numbers correctly.

	Blank	Segment / Port Pin							Result
Digit	MSB	g / PA6	f / PA5	e / PA4	d / PA3	c / PA2	b / PA1	a / PA0	
0	0	0	1	1	1	1	1	1	**0x3F**
1	0	0	0	0	0	1	1	0	**0x06**
2	0	1	0	1	1	0	1	1	**0x5B**
3	0	1	0	0	1	1	1	1	**0x4F**
4	0	1	1	0	0	1	1	0	**0x66**
5	0	1	1	0	1	1	0	1	**0x6D**
6	0	1	1	1	1	1	0	1	**0x7D**
7	0	0	0	0	0	1	1	1	**0x07**
8	0	1	1	1	1	1	1	1	**0x7F**
9	0	1	1	0	1	1	1	1	**0x6F**

Table 5.1: Numbers to Write to the 7-segment Display on the Microcontroller Port

What the software must do to drive a single seven-segment digit

To drive the single display configurations shown in **figure 5.6**, the software must:

1. Initialise the microcontroller port pins interfaced to the display as outputs.
2. Allow writing to the port pins via a register.
3. Write the values from **table 5.1** to the port using a lookup table to display each digit.

The **`seven_seg_single_RAM`** example program configures port pins PA0 to PA7 as outputs that can be controlled by writing to a microcontroller register. Port pin PA7 is wired to the decimal point of the seven-segment display, but is not used in the program. A **`for`** loop in the program counts through the numbers 0 to 9. Array **`g_disp_table[]`** is used as a look up table to look up the value to write to the port to display the numbers that the **`for`** loop variable **`index`** is counting through.

Driving more than one seven-segment display

The circuits shown so far will only drive a single seven-segment display. To drive a second display, another eight port pins could be used. This method would allow only four seven-segment displays to be controlled by a single microcontroller with 32 I/O pins and would not leave spare pins to interface to any other devices. To solve this problem, we can multiplex the displays.

Figure 5.8 shows two seven-segment displays multiplexed and driven by transistors. The displays share a common data bus for driving the individual display segments. The common anodes of each display are each controlled by a dedicated microcontroller pin. Only one display is switched on at a time by supplying power to its common anode. Driving the common data bus with transistors allows the LEDs to be driven with higher currents.

Principle of operation

The idea is to switch the first display on, and then display the desired number on it, switch it off, switch the second display on and display the desired number on it. When this is done fast enough, the displays appear to be permanently on and can each display a different number. With two seven-segment displays, we can now display decimal numbers up to a value of 99.

Adding more seven-segment displays

To add more seven-segment displays to the circuit of **figure 5.8,** an additional microcontroller pin is required per display and is wired with an additional two transistors as already shown. The same data bus for driving the segments is used for each display.

The software must now switch between the additional displays. As more displays are added, there is less "on" time for each display which will cause the displays to be dimmer. To increase the brightness of the displays, they can be driven with higher current. Driving the display segments with transistors allows driving them with higher currents by decreasing the resistor values on the transistor collectors.

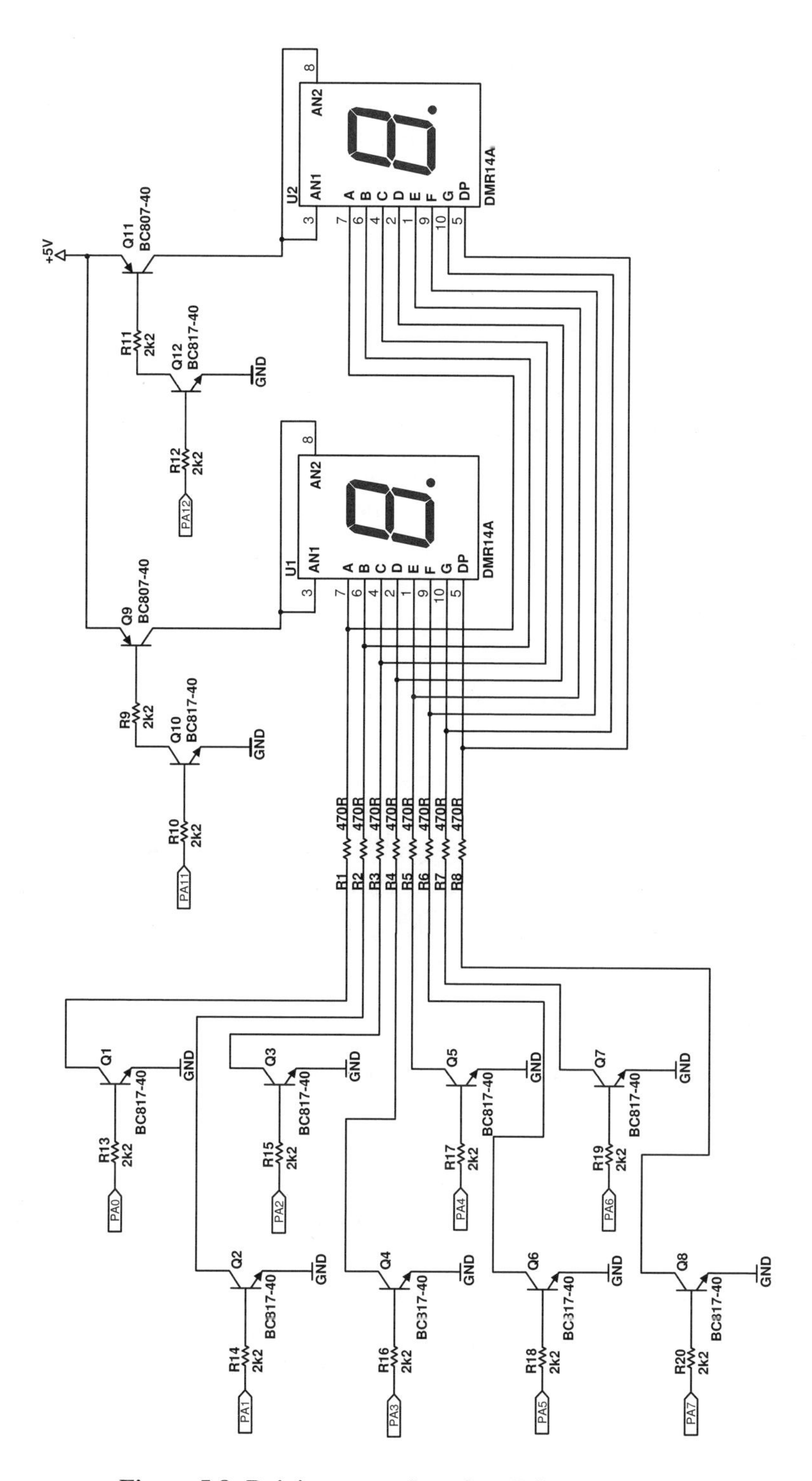

Figure 5.8: Driving more than One 7-Segment Digit

Software for driving two displays

The **seven_seg_2_RAM** program drives the circuit of **figure 5.8**. This program uses interrupts to switch the correct seven-segment display on at the correct time. It initialises the following:

1. Sets up PA0 to PA7 as outputs to drive the segments of the display.
2. Sets up an additional two pins (PA11 and PA12) as outputs to switch the individual displays on and off.
3. Enables timer/counter 0 (TC0) interrupt.
4. Sets up TC0 to generate an interrupt every 10mS.

After initialisation, the main program loop runs that contains a **for** loop that counts from 0 to 99. Every 10mS when TC0 generates an interrupt, the ISR will switch on the next seven-segment display and display the part of the number that must be displayed on it.

Each element of the global array **g_seg_num[]** holds the number for its corresponding seven-segment display. The number that the array element holds will index the look up table and switch on the correct segments of the display that the ISR is currently switching on.

The **Num2Seg()** function called in the **for** loop converts the **count** number to units and tens that will be stored in the global array **g_seg_num[]**. If additional seven-segment displays were to be added to the circuit, the **Num2Seg()** function would also need to get the number of 100s and 1000s, etc. to display on the additional displays.

Hardware Tip

If you design a PCB with four seven-segment displays and the centre pins of each set of pins on the display are the common pins (pins 3 and 8 on the SunLed display used in this chapter), you can solder the third display from the left upside-down. The upside-down display will need its own lookup table in software, but the four displays can now be used as a digital clock as shown in **figure 5.9**. Not all seven-segment displays can be used like this as they are not all symmetrical, so always check the datasheet for the display that you are using. The same PCB can now be used as a clock or a four digit number with decimal points between each number, depending on which way the third digit from the right is soldered.

Figure 5.9: The Decimal Points of the Middle Two Displays Make it into a Digital Clock

Consecutive and non-consecutive pins – software considerations

It may not be possible to have all of the data bus pins from the microcontroller in consecutive order as shown in this chapter. If the pins are in consecutive order but not starting from PA0, software can shift the data to be written to the device to the left. For example if microcontroller pins PA15 to PA22 are used as an 8-bit data bus to drive seven segment displays or an LCD display, the software programs shown in this chapter could be used by only changing one thing – shift the data that is to be written to the device to the left by 15 before writing it. Just make sure that it is in a 32 bit word before shifting it.

When pins used for the data bus are scattered around the microcontroller's 32-bit port, there are two approaches that can be take. Either use a lookup table or shift the individual bits or groups of bits to their correct positions before writing them to the device. In the seven segment display examples, it is easy to change the lookup table. The current lookup table will need to be changed to 32-bit words so that any pin on the microcontroller pin could be written to. The correct data would then need to be entered into the lookup table to display the correct digits. This works well because of the mask that is used when setting up the microcontroller port as an output prevents any of the other port pins from being affected by a write. Of course this relies on there being only one peripheral device being accessed in this manner. If there were two devices that needed to be written to by this method, the bits in the word to be written that belong to the other device would need to be left unchanged by first reading it and then modifying the data to be written to first device to include it.

A lookup table for an LCD display may not be practical. For an 8-bit bus, the table would need to be 256 bytes long. It is better to shift the data to be written to the display to its correct position before writing.

6. DC Motors and Stepper Motors

6.1 DC Motors

Chapter 2 showed how to interface a DC motor to a microcontroller using a transistor to switch the motor on and off only. This section will use **Pulse Width Modulation (PWM)** to control the speed of the motor and an **H-bridge** to control the direction of the motor (forward or reverse).

6.1.1 DC Motor Speed Control

Any of the circuits from **figure 2.17** to **figure 2.20** can be used with the PWM although the optocoupler used in the last circuit can only be used at low PWM frequencies. The idea is to connect the transistor to one of the microcontroller pins that can internally be connected to the PWM.

On the AT91SAM7S256, the PWM channels can be internally connected to pins PA0 to PA2, PA11 to PA14 or PA23 to PA25. This information is obtained from **table 10-3** in the AT91SAM7S datasheet.
The simple circuit of **figure 6.1** was used to speed control a 12V DC, 0.08A fan motor using PWM0 on pin PA0. PWM0 is channel 0 of the AT91SAM7S internal PWM controller.

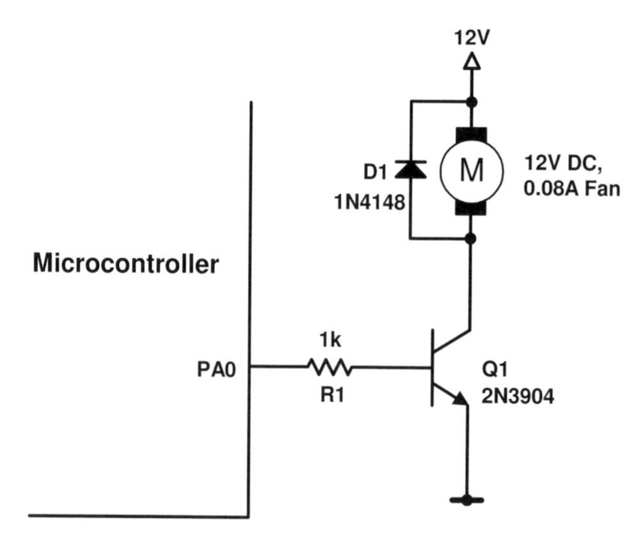

Figure 6.1: DC Motor Interfaced to PWM0 on Pin PA0 for PWM Control

How Pulse Width Modulation (PWM) Control Works:
Pulse Width Modulation refers to the production of a square wave pulse at a fixed frequency and then varying the width of the pulses in order to control the output voltage level. Changing the pulse width changes the duty cycle of the output pulses.

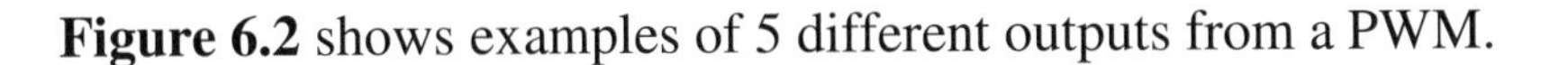

Figure 6.2 shows examples of 5 different outputs from a PWM.

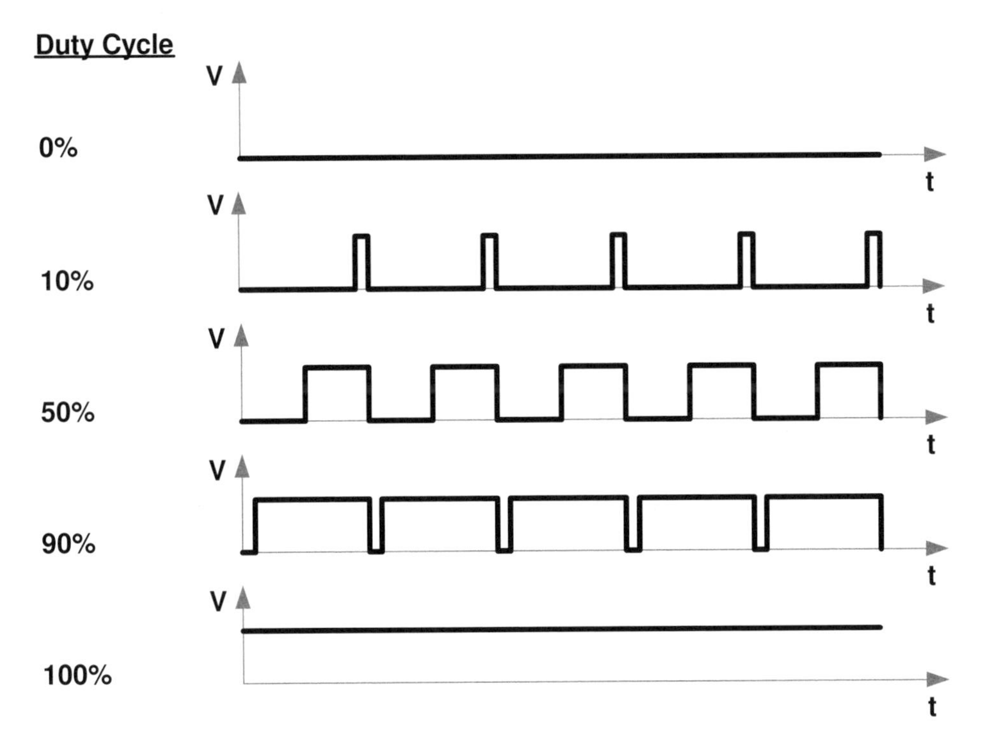

Figure 6.2: Examples of Pulse Width Modulation Outputs with Various Duty Cycles

When a PWM output has a duty cycle of 0%, the output voltage will be 0V.

When the PWM output duty cycle is 10%, and driving a maximum of 12V, the effective voltage will be 10% of 12V i.e. 1.2V.

When the PWM duty cycle is at 50%, the 12V pulses appear as 50% of 12V, i.e. 6V, etc.

When the PWM duty cycle is 100%, the PWM output will be 12V in this example.

If an internal PWM channel of the AT91SAM7S is configured and driven on the microcontroller pin, then the voltage measured on the pin will be between 0% and 100 % of 3.3V. E.g. if the duty cycle is 75%, the voltage measure on a DC voltmeter will be 75% of 3.3V = 2.475V.
The `dc_motor_PWM_RAM` program sets up PWM channel 0 (PWM0) on pin PA0 of the AT91SAM7S256 to drive the fan motor circuit of **figure 6.1**. The program interfaces to a terminal emulator program running on the PC. The 'u' and 'd' keys are pressed to increase or decrease the duty cycle which in turn increases or decreases the fans speed **Figure 6.3** shows how the program sets up the PWM channel. The duty cycle and equivalent DC voltage of the PWM output are calculated and displayed on the terminal emulator.

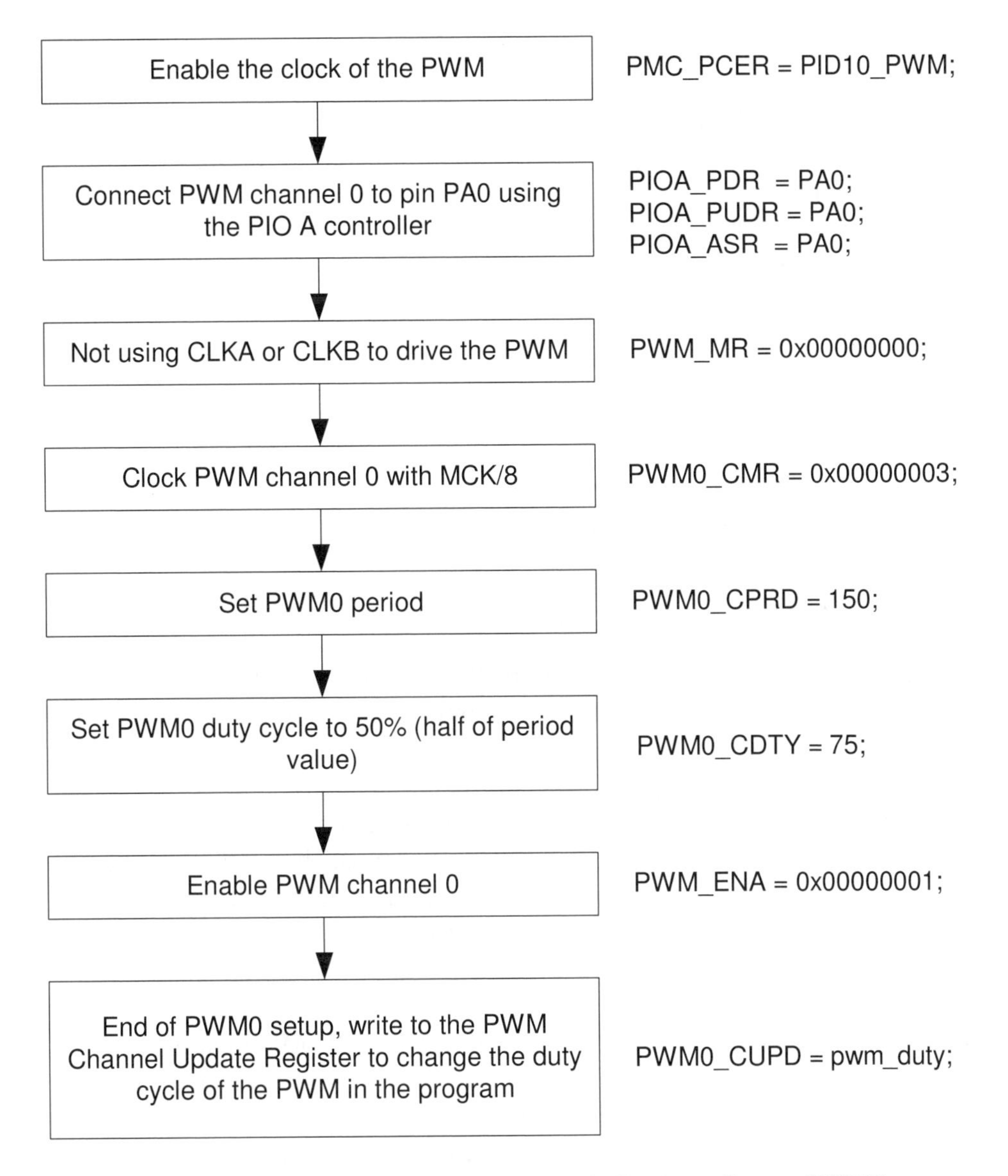

Figure 6.3: How the `dc_motor_PWM_RAM` Program Sets up PWM0

The frequency of PWM 0 in the example program is calculated as follows:
The PWM0_CMR register was set up to supply PWM0 with a clock frequency of MCK/8. MCK is the master clock and is set to 47.9232MHz if using an 18.432MHz crystal and the template files supplied with this book.
MCK/8 = 5.9904MHz

The PWM0 period was set in the PWM0_CPRD register to a count of 150.

The period of the clock supplied to PWM0 is:
clock period = 1/5.9904MHz
= 166.934ns
The period of the output waveform will be the time taken to count to 150 at the input clock frequency:
PWM0 period = 166.934ns × 150
= 25.04μs

PWM0 frequency will then be:
1/25.04μs = 39.936kHz
OR
PWM frequency = input clock / CPRD register value
= 5.9904MHz / 150
= 39.936kHz

6.1.1.1 Silicon Errata

When a bug is discovered in a microcontroller (in the actual silicon) after it has gone into production, the microcontroller manufacturer will produce an errata document that provides a solution or workaround to the problem. Atmel includes the microcontroller errata as part of its datasheet – in chapter 40 of the AT91SAM7S datasheet. The bug may be fixed in a later revision of the microcontroller, so the user of the microcontroller must first check the date of manufacture and/or silicon revision number to determine which errata apply to the particular microcontroller being used. This information is provided in the datasheet.

In the case of the PWM controller, the polarity of the PWM signal may be changed if 0 or 1 is written to the PWM_CDTY register. To prevent this problem, 0 or 1 must not be written to the PWM_CDTY register. As the PWM_CUPD register is used as a double-buffer to update the PWM_CDTY register, 0 or 1 must not be written to the PWM_CUPD register.

6.1.2 DC Motor Direction Control

An H-Bridge configuration is commonly used to control the direction of a DC motor. **Figure 6.4** shows an example of an H-Bridge circuit using bipolar transistors. The circuit was designed for low current 5V DC motors.
How an H-Bridge works:

In **figure 6.4**, Q1 and Q4 are switched on at the same time to make the motor run in one direction. Q2 and Q3 are switched on to make the motor run in the opposite direction.

Q1 and Q3 must never be switched on at the same time or they will short out the 5V supply rails. The same goes for Q2 and Q4. IC U1 provides the logic that prevents the combination of transistors switching on that would short out the supply rails. The table in the figure displays the possible logic combinations from two microcontroller pins. The motor can only move forward or in reverse or be switched off.

It is possible to connect Q1, Q2, Q3 and Q4 directly to the microcontroller, eliminating two transistors and the IC from the circuit, but any error in the program used to control this circuit could result in the wrong combination of transistors switching on and being destroyed.

Although the direction of the motor is shown in the figure, the direction depends on which wire from the motor is connected to which side of the H-Bridge.

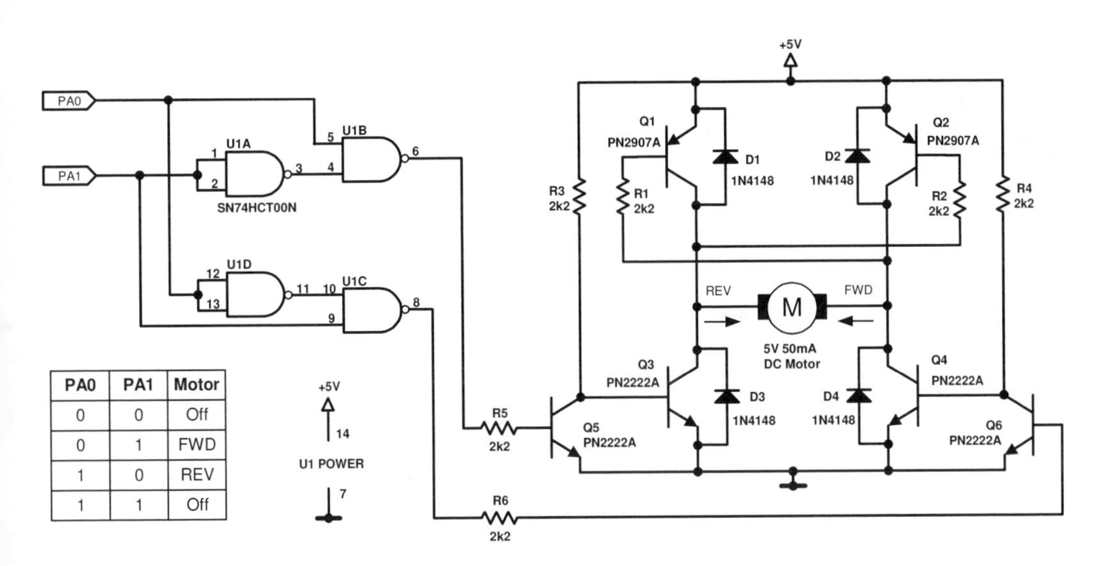

PA0	PA1	Motor
0	0	Off
0	1	FWD
1	0	REV
1	1	Off

Figure 6.4: An H-Bridge Circuit for a Low Current 5V DC Motor

Figure 6.5 shows a similar circuit to **figure 6.4**, but can drive a bigger motor. The circuit also allows for braking of the motor by switching on Q3 and Q4 at the same time which grounds the wires to the motor.

The circuit uses PNP and NPN Darlington transistors to provide the extra power to the bigger motor. **Heat sinks** must be attached to these transistors in order to dissipate the heat generated in them.

According to the TIP122/127 datasheet, these transistors can have a V_{CE} of up to 2V, which would only deliver 8V to the motor from a 12V power supply. The transistors that were used when testing this circuit weren't that bad, but still reduced the voltage supplied to the motor. Because the voltage to the motor was reduced, the current drawn by the motor was also reduced.

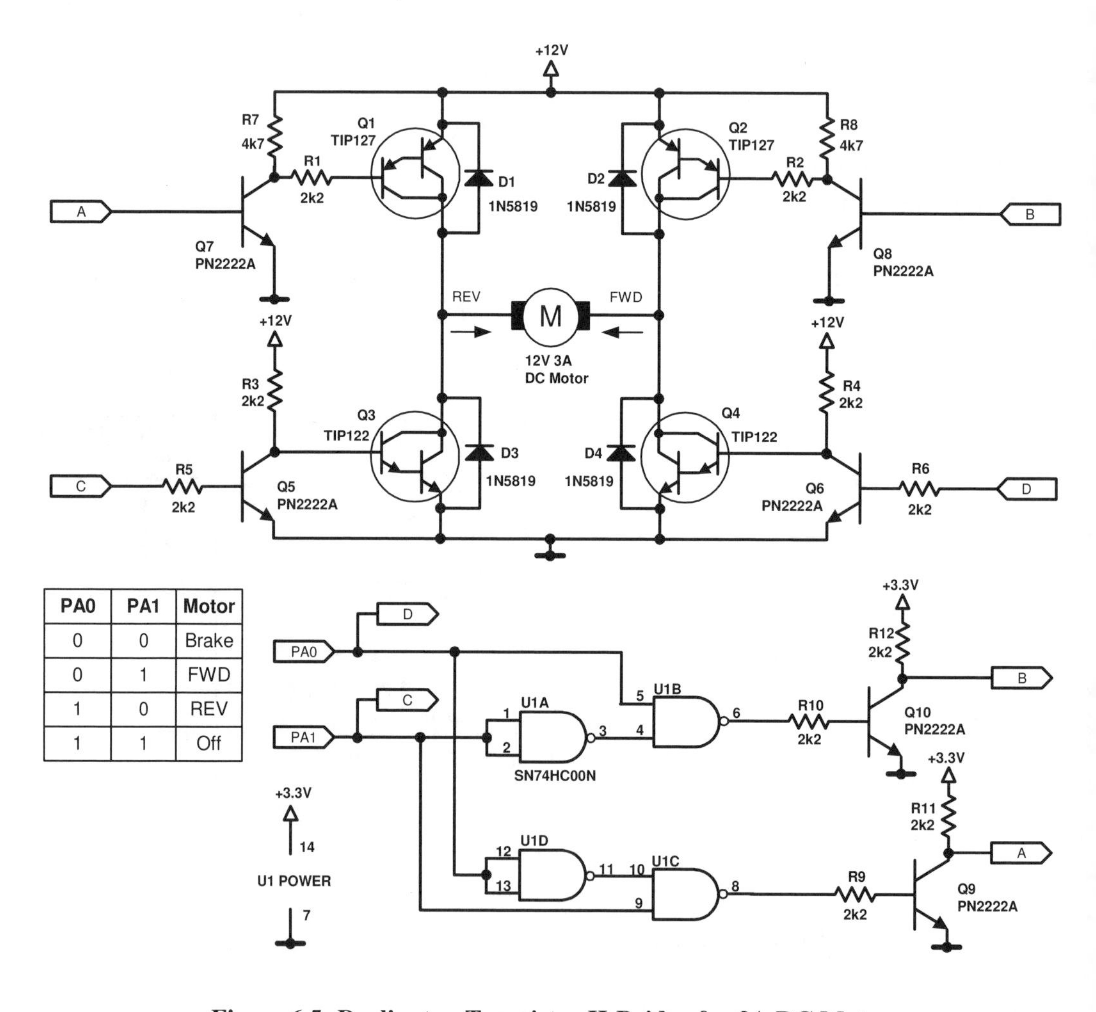

PA0	PA1	Motor
0	0	Brake
0	1	FWD
1	0	REV
1	1	Off

Figure 6.5: Darlington Transistor H-Bridge for 3A DC Motor

Software to drive the H-Bridge circuits of figures 6.4 and 6.5:
The **dc_motor_H_bridge_RAM** program operates the H-Bridges without speed control, simply switching the motor on for a period of time, then off, and then on in the opposite direction.

The **dc_motor_H_bridge_PWM_RAM** program uses PWM to control the H-Bridge. The program connects PWM 0 to pin PA0 and PWM 1 to PA1. To operate the motor in one direction, one of the PWMs is switched fully off and the other controls the speed using PWM. To operate the motor in the opposite direction, the opposite PWM is switched fully off and the other PWM is used to control the speed.
The program ramps the motor speed up by increasing the PWM duty cycle from 0% to 100%, lets it run at a constant speed for a while, then ramps the speed down and does the same in the reverse direction.

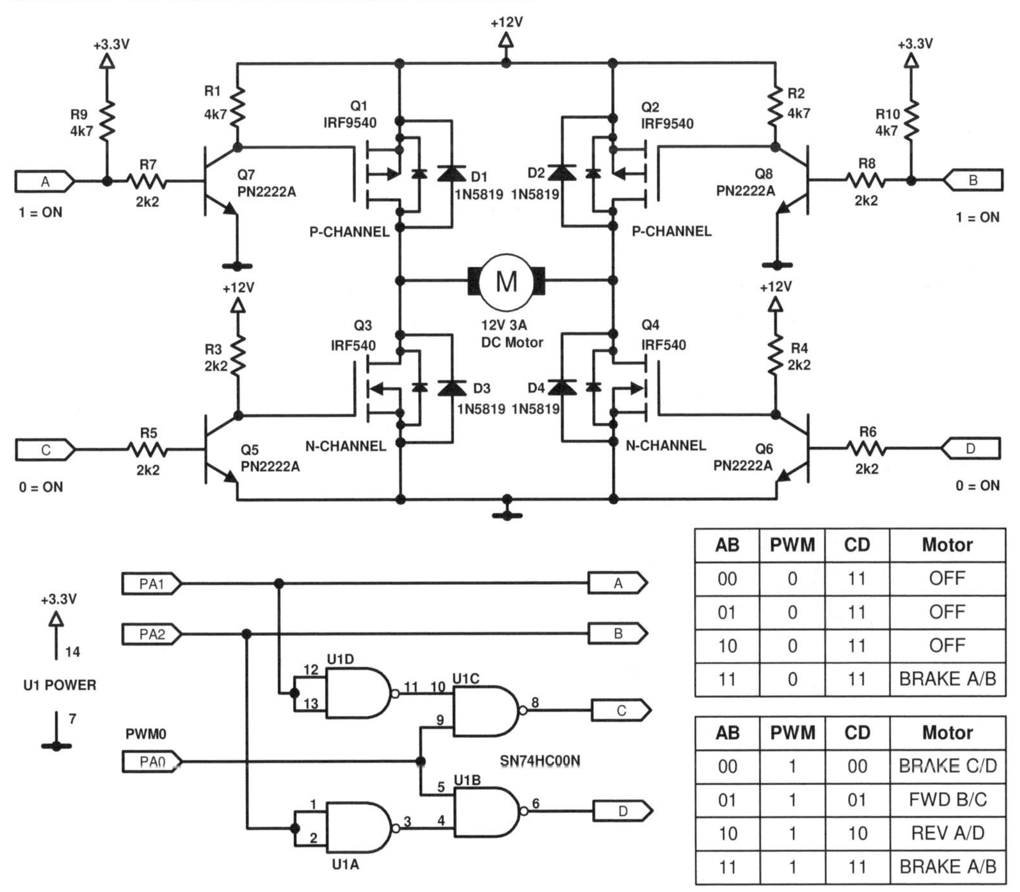

AB	PWM	CD	Motor
00	0	11	OFF
01	0	11	OFF
10	0	11	OFF
11	0	11	BRAKE A/B

AB	PWM	CD	Motor
00	1	00	BRAKE C/D
01	1	01	FWD B/C
10	1	10	REV A/D
11	1	11	BRAKE A/B

Figure 6.6: MOSFET H-Bridge for 3A DC Motor

Figure 6.6 shows an H-Bridge that uses MOSFETs. This circuit also uses a logic protection circuit to prevent the wrong combination of transistors from switching on. It uses three microcontroller pins to control the H-Bridge. Two are used for switching the bridge and the third is used for PWM control.

This circuit can switch the motor off, brake the motor by switching Q1 and Q2 on, brake the motor by switching Q3 and Q4 on, PWM control the motor in the forward direction and PWM control the motor in the reverse direction.
Software to drive the MOSFET H-Bridge:
The **`dc_motor_MOSFET_H_Bridge_PWM_RAM`** program controls the H-Bridge of **figure 6.6**. The program allows the H-Bridge to be controlled via a terminal emulator. Motor speed and direction can be controlled by sending various characters from the terminal emulator to the AT91SAM7S DBGU port. A menu of commands is displayed on the terminal emulator when the program is started.

6.2 Stepper Motors

Two main types of permanent magnet stepper motors are:

Unipolar Stepper Motors – Use windings with a centre-tap.
Bipolar Stepper Motors – Use windings with no common connection.

We will cover **unipolar** stepper motors in this section. Bipolar stepper motors require an H-bridge as the current must be reversed through the windings to operate the motor.

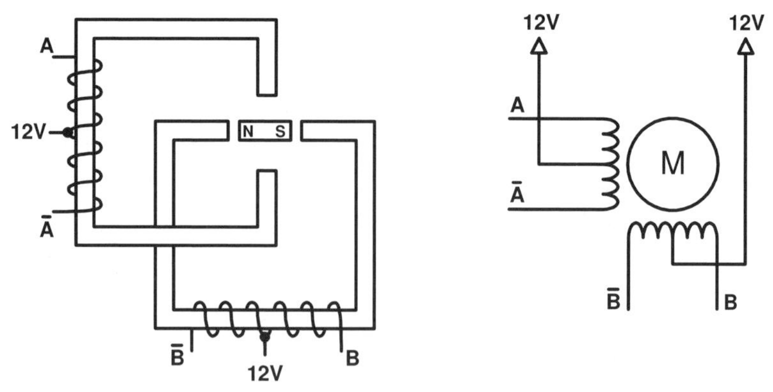

Figure 6.7: Simple Stepper Motor Construction (left) and Stepper Motor Schematic Symbol (right)

Figure 6.7 shows the principle of operation of a simple stepper motor. The permanent magnet rotor is situated in the centre of two pairs of electromagnetic poles. This motor will produce four steps per revolution as the permanent magnet rotor is rotated between the electromagnetic poles. Practical motors will have many more steps per revolution.

To rotate the motor, the electromagnetic poles are activated in a sequence by grounding the terminals $\overline{A}$, A, B or B. The motor direction is reversed by reversing the sequence.

Figure 6.8 shows various modes that are used to drive a stepper motor.

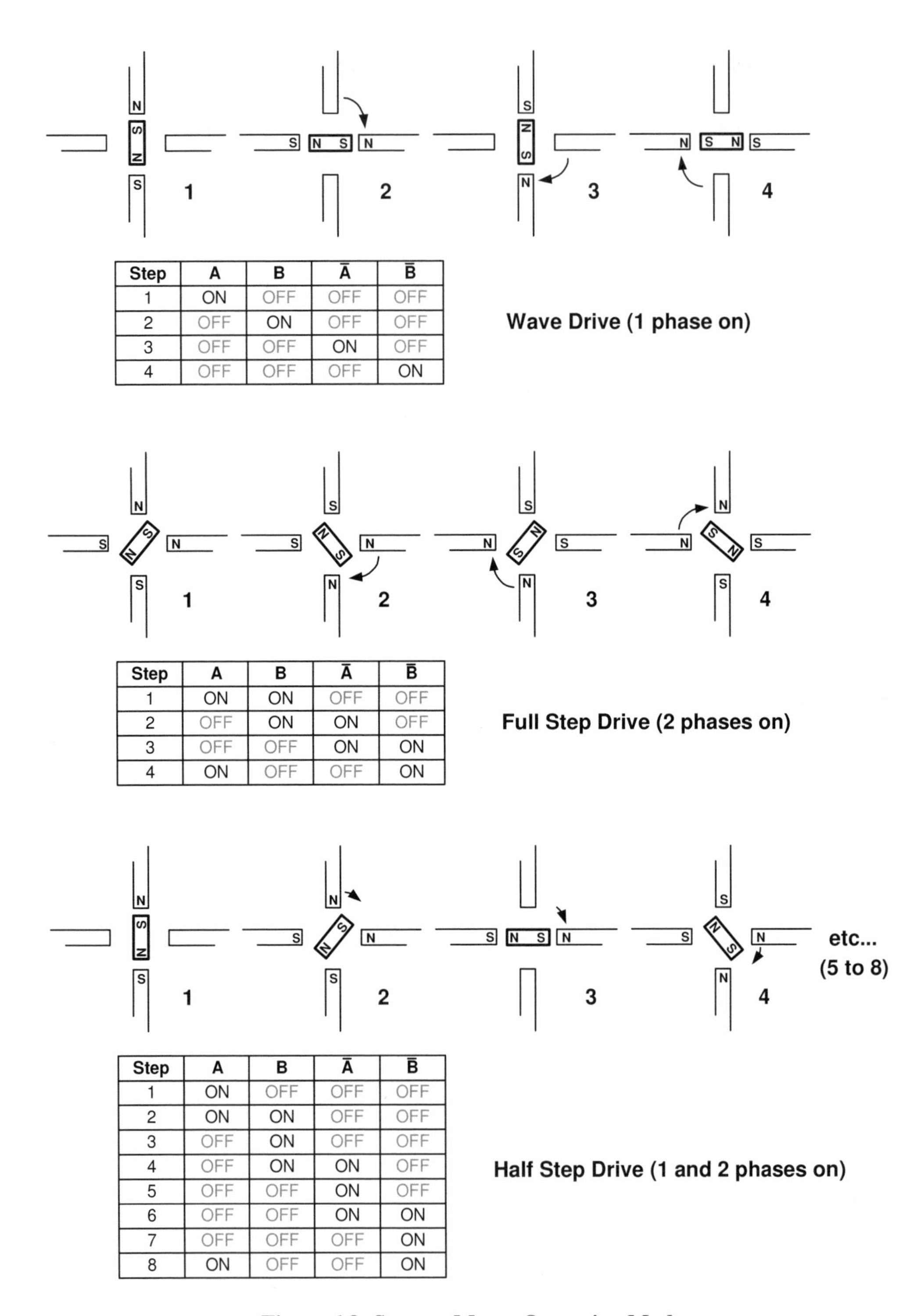

Step	A	B	$\bar{A}$	$\bar{B}$
1	ON	OFF	OFF	OFF
2	OFF	ON	OFF	OFF
3	OFF	OFF	ON	OFF
4	OFF	OFF	OFF	ON

Step	A	B	$\bar{A}$	$\bar{B}$
1	ON	ON	OFF	OFF
2	OFF	ON	ON	OFF
3	OFF	OFF	ON	ON
4	ON	OFF	OFF	ON

Step	A	B	$\bar{A}$	$\bar{B}$
1	ON	OFF	OFF	OFF
2	ON	ON	OFF	OFF
3	OFF	ON	OFF	OFF
4	OFF	ON	ON	OFF
5	OFF	OFF	ON	OFF
6	OFF	OFF	ON	ON
7	OFF	OFF	OFF	ON
8	ON	OFF	OFF	ON

Figure 6.8: Stepper Motor Operating Modes

6.2.1 Wave Drive

The top illustration in **figure 6.8** shows how a stepper motor is operated in wave drive mode. The table shows the sequence of windings to activate in order to rotate the motor. Wave drive mode moves the rotor to align the poles of its permanent magnet with the electromagnetic poles. A stepper motor with more poles will be operated in exactly the same way as this simple motor, with the exception that each step of the wave drive sequence will only move the rotor of the motor a small fraction of a complete turn. To rotate a stepper motor that has more poles a full revolution, the windings will need to be activated in the sequence many times.

A unipolar stepper motor driven in wave drive mode uses only 25% of the windings which means that maximum torque of the motor is not being utilised.

6.2.2 Full Step Drive

Full step drive mode is shown in the centre of **figure 6.8**. Two windings (phases) are energised for each step in full step drive mode.

A unipolar motor driven in full step drive mode uses 50% of the windings which increases the torque of the motor compared to wave drive mode.

6.2.3 Half Step Drive

Half step drive mode combines the steps of wave drive and full step drive mode. The stepper motor will run more smoothly in half step mode as the stepper motor will now take more steps per revolution.

6.2.4 Microstepping Drive

Microstepping drive refers to continuously varying the currents in the windings in order to break a single step of motor into several smaller steps. Microstepping is an advanced technique and will not be covered in this book.

Figure 6.9 shows a practical stepper motor interface circuit that uses opto-isolation and MOSFETs to drive the motor from an AT91SAM7S microcontroller.

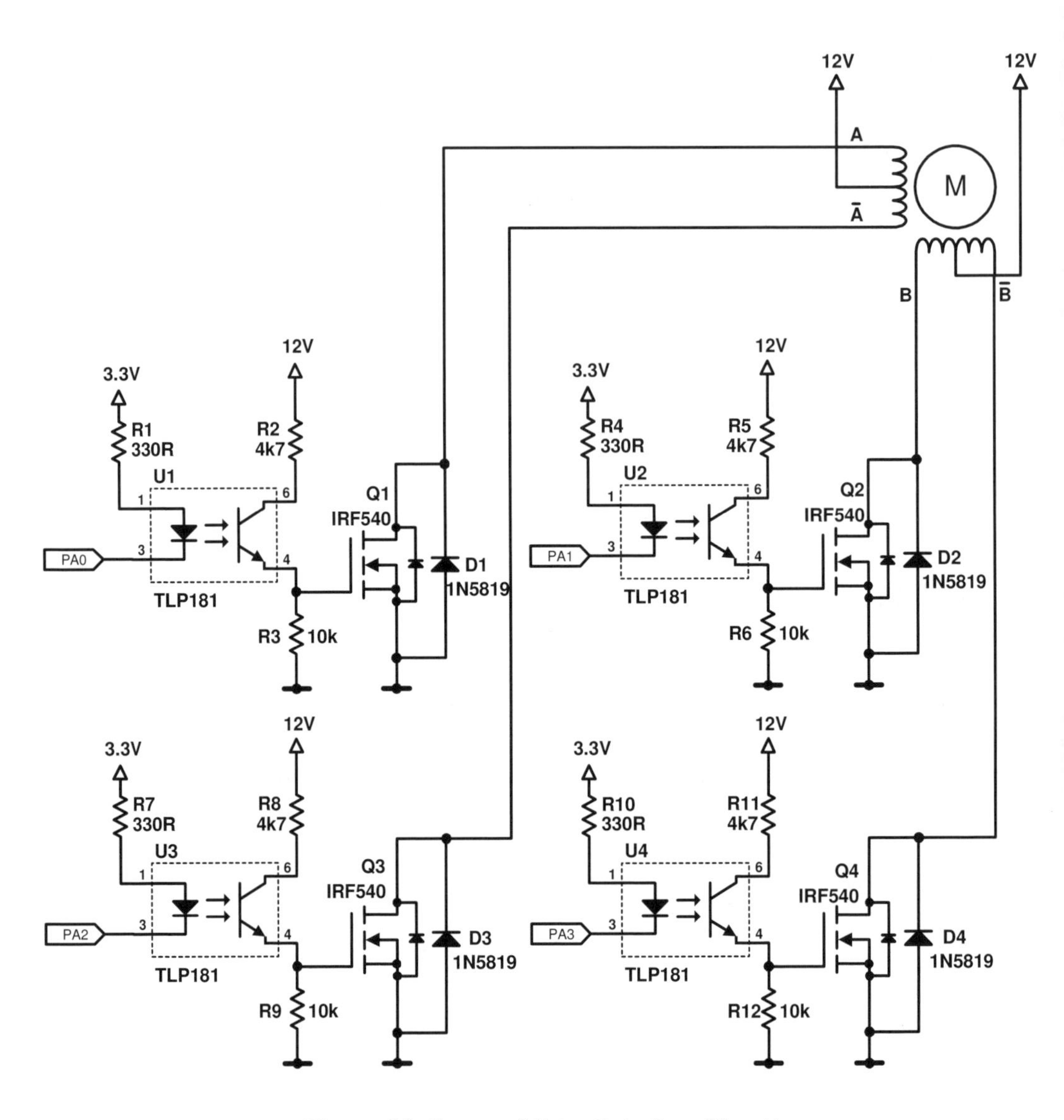

Figure 6.9: Stepper Motor Interface Circuit

Software to drive the stepper motor circuit:

The `stepper_motor_RAM` program drives the stepper motor circuit of **figure 6.9** in half step drive mode. The program connects to a terminal emulator on the PC allowing the user to increase or decrease the speed of the motor, drive the motor in forward or reverse direction and stop the motor.

Speed of the motor is controlled by a timer that regulates the speed at which the stepper motor control sequence is changed.

7. Analogue Signals and PWM Audio

7.1 ADC

An Analogue to Digital Converter (ADC) converts an analogue voltage applied to one of the microcontroller ADC pins to a number that the microcontroller can use. The analogue voltage that the ADC can measure can be of a value between 0V and a reference voltage that is applied to a reference voltage pin.

Analogue = British English
Analog = American English

AT91SAM7S microcontrollers have an **8-channel** Successive Approximation Register (SAR) Analogue to Digital Converter (ADC) capable of being operated in **10-bit** or **8-bit** resolution modes. ADC pins AD4 to AD7 are dedicated for use with the ADC. AD0 to AD3 are multiplexed on I/O pins.

Table 7.1 shows the operating speeds of the ADC in 10-bit and 8-bit modes.

ADC Clock / Resolution	Conversion Time	Throughput
5MHz / 10-bit	2μS	384k Samples per Second
8 MHz / 8-bit	1.25μS	533k Samples per Second

Table 7.1: AT91SAM7S ADC Specification

7.1.1 ADC Pins and Reference Voltage

Figure 7.1 shows the AT91SAM7S ADC pins as well as various ways of providing a reference voltage to the ADC on the ADVREF pin. The reference voltage for the ADC can be in the range of **2.6 to 3.3V**. The input voltage on the ADC pins being measured can be between 0V and ADVREF.

Reference voltage from power supply

The circuit on the left side of **figure 7.1** uses the I/O voltage of the microcontroller as a reference voltage. The accuracy of this reference depends on the accuracy of the 3.3V regulator supplying the microcontroller and may be susceptible to noise on the power supply. The decoupling capacitor is an attempt to remove some noise.

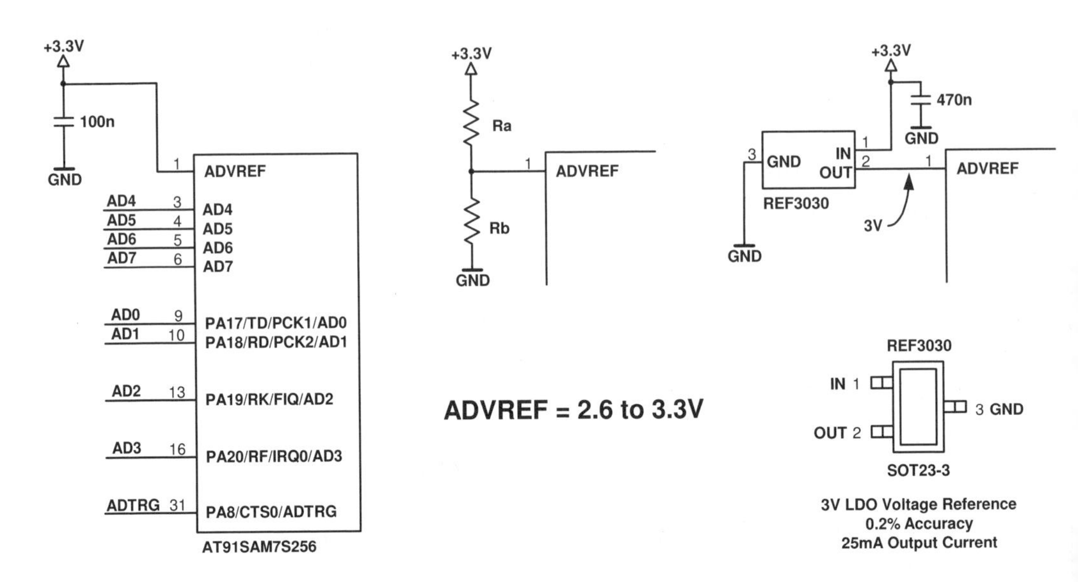

Figure 7.1 ADC Pins on the AT91SAM7S and Various Reference Voltage Options

Reference voltage from voltage divider or zener diode

The middle circuit of the figure uses a voltage divider to provide a reference voltage to the ADC. This method is not accurate at all. Firstly it depends on the tolerance of the resistors and secondly as more ADC channels are enabled, the current drawn by the ADVREF pin increases, changing the reference voltage. An improvement on this circuit would be to change resistor Rb for a zener diode. Be aware that most common zener diodes typically have a tolerance of 5%, although more accurate zeners are available. Alternatively a unity gain op-amp buffer can be used to buffer the voltage divider (as shown in **figure 7.2** used for buffering an ADC input in this figure).

Some measurements were taken on the voltage divider reference circuit as follows: With Ra = 220Ω and Rb = 6k8 the reference voltage should be 3.1967V.

The following measurements were taken with the specified channels enabled, after ADC conversion was started:
At power up, no program running: **3.19V**
AD4 enabled: **3.16V**
AD4 & AD5 enabled: **3.15V**
AD4, AD5 & AD6 enabled: **3.15V**
AD4, AD5, AD6 & AD7 enabled: **3.15V**

Reference voltage from IC
A low drop out (LDO) voltage reference IC to supply a 3V reference to the ADC is shown on the right of **figure 7.1**. There are a number of different voltage reference ICs to choose from, this is just one example. If an accurate 3.3V reference is required, the REF3033 could be used but would have to be supplied with a voltage greater than 3.3V, e.g. 5V.

7.1.2 Basic ADC Interfacing

Figure 7.2 shows two methods of interfacing a potentiometer to the ADC of the AT91SAM7S so that you can start experimenting with the ADC.

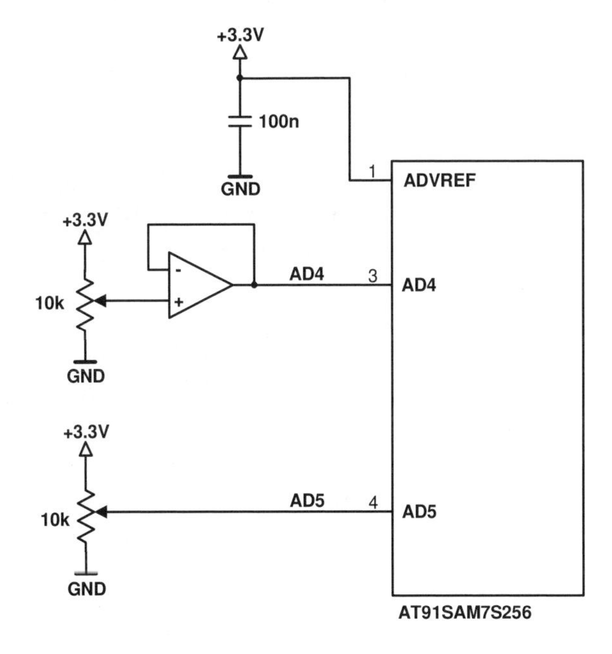

Figure 7.2: ADC used to Measure a Voltage on a Potentiometer

Direct connection

A potentiometer is connected directly to AD5 in the figure. This will enable the voltage on AD5 to be varied between 0V and 3.3V. The ADC will convert these voltages to a value between 0 and 1023 when it is operated in 10-bit mode (0 to 255 in 8-bit mode). If a 3V reference voltage was used on the ADVREF pin, the voltage to be measured on AD5 would have to be limited to between 0 and 3V. 3V on AD5 would then be converted to 1023 by the ADC.

Unity gain buffer (op-amp)

An op-amp is used on AD4 to buffer the input voltage. The op-amp is configured as a unity gain buffer (gain of one). This circuit provides current gain and can be used when the voltage being measured by the ADC can not supply high enough current to drive the ADC circuit.

Typically it would be desired to use an op-amp that runs from a single power supply (i.e. no negative voltage is required to power it) and is low voltage so that it can operate from the 3.3V supply of the microcontroller. Another good feature of the op-amp would be if it were a rail-to-rail op-amp. This means that the output voltage would be able swing within a few mill-volts of the power supply rails. When using an op-amp that is powered from the microcontrollers 3.3V supply, the output voltage will never be able to swing exactly to 0V or to 3.3V, so the full range of the ADC will not be used.

Op-amp examples

Some examples of op-amps are shown in **table 7.2**. All the op-amps in the table are quad op-amps (4 op-amps in each IC) available in 14 pin packages. These are just some op-amps that are available; there are many more op-amps from a number of different vendors.

	Supply Range	**+V Rail Swing**	**0V Rail Swing**	**ADC Range (10-bit mode)**	**Comments**
LM324	3V to 32V	2.07V	0V	0 to 641	Older IC SMT and DIP
MC33204	1.8V to 12V	3.27V	0.04V	11 to 1019	Rail-to-rail SMT and DIP
AD8040	2.7V to 12V	3.29V	0.02V	5 to 1022	Rail-to-rail SMT only

Table 7.2: Examples of Op-amps

Some of the op-amps listed are available in packages with dual or single op-amps:
MC33204 quad op-amp
MC33202 dual op-amp
MC33201 single op-amp

AD8040 quad op-amp
AD8030 dual op-amp
AD8029 single op-amp

Simple ADC software
The **`ADC_DBGU_port`** program can be used with either of the configurations in **figure 7.2** to read back a value from the dedicated ADC channels on pins AD4 to AD7. The program displays the ADC channel values on a terminal emulator connected to the DBGU port of the AT91SAM7S.

At start-up the program resets the ADC by setting the SWRST bit in the ADC_CR:
```
ADC_CR = 0x00000001;
```

The ADC_MR is written to as follows:
```
ADC_MR = 0x0F1F3F00;
```

This disables hardware triggers, selects 10-bit resolution, puts the ADC into normal mode and sets up the following timing parameters from bit fields in the ADC_MR:

PRESCAL = 0x3F (63)
With MCK of 47.9232MHz, this sets the ADCClock to **374.4kHz**

STARTUP = 0x1F (31)
With an ADCClock of 374.4kHz, this sets Startup Time to **683.761mS**

SHTIM = 0xF (15)
With an ADCClock of 374.4kHz, this sets up Sample & Hold Time to **40.064µS**

```
ADC_CHER = 0x000000F0;
```
Enables ADC channels 4 to 7

```
ADC_CR = 0x00000002;
```
Starts the ADC conversion

Inside the **while(1)** loop, the ADC_SR is checked to see if all four ADC values have been converted from analogue to digital:

```
if (ADC_SR & 0x000000F0) {
```

If all four values have been converted, the values are read from their corresponding Channel Data Registers (ADC_CDRx) and sent to a terminal emulator on the DBGU port. The ADC is then restarted to perform the next conversion, this causes the program to continually convert and display the values that the ADC is reading.

This program only displays the digital value that the ADC has converted from the analogue value read. The next section shows how to convert the ADC code to the voltage that is being sampled (read) on the ADC pin. This is related to the voltage reference talked about earlier.

Converting the ADC Code to a Voltage

Each unit of the ADC code that is read represents one measure of the ADC reference voltage – the reference voltage divided by the number of samples that the ADC can take over this reference range. The number of samples (or resolution) depends on the number of ADC bits that the ADC code consists of e.g. 256 samples for an 8-bit ADC or 1024 samples for a 10-bit ADC. Mathematically this is represented as follows:

For an **8-bit ADC** with **3.3V reference**:

$$
\begin{aligned}
N_{SAMPLES} &= 2^N \\
&= 2^8 \\
&= 256
\end{aligned}
$$

Voltage resolution:

$$
\begin{aligned}
V_{RESOLUTION} &= \frac{V_{REF}}{N_{SAMPLES}} \\
&= \frac{3.3}{256} \\
&= 12.891mV
\end{aligned}
$$

For a **10-bit ADC** with **3.3V** reference:

$$\begin{aligned} N_{SAMPLES} &= 2^N \\ &= 2^{10} \\ &= 1024 \end{aligned}$$

Voltage resolution:

$$\begin{aligned} V_{RESOLUTION} &= \frac{V_{REF}}{N_{SAMPLES}} \\ &= \frac{3.3}{1024} \\ &= 3.223mV \end{aligned}$$

From this we can see that each unit of an 8-bit ADC code has the value of 12.891mV and a 10-bit ADC has the value of 3.223mV.

A program running on the microcontroller can calculate the analogue voltage as follows:

$$V_{ADC} = \frac{V_{REF}}{2^N - 1} \times ADCCODE$$

Where:
V_{ADC} = Analogue voltage on ADC pin (voltage that we want to measure)
V_{REF} = ADC reference voltage (on microcontroller ADC reference pin)
N = ADC resolution (e.g. 8-bit or 10-bit)
ADCCODE = value that the microcontroller reads from the ADC register

Example:
A program running on the microcontroller reads 830 from the ADC conversion register on a 10-bit ADC with a 3.3V reference voltage.

$$\begin{aligned} V_{ADC} &= \frac{V_{REF}}{2^N - 1} \times ADCCODE \\ &= \frac{3.3}{2^{10} - 1} \times 830 \\ &= \frac{3.3}{1024 - 1} \times 830 \\ &= 2.677V \end{aligned}$$

I.e. the voltage being read on the ADC pin is 2.677V

For a 10-bit ADC the formula can be simplified to:

$$V_{ADC} = \frac{V_{REF}}{1023} \times ADCCODE$$

Some texts also use the formula:

$$V_{ADC} = \frac{V_{REF}}{2^N} \times ADCCODE$$

For a 10-bit ADC this can be simplified to:

$$V_{ADC} = \frac{V_{REF}}{1024} \times ADCCODE$$

This is also correct, but it will never display the maximum voltage that the ADC can read i.e. a voltage that equals the reference voltage. This is because the highest value that can be read from the ADC data register is 1023 for 10-bit ADC (1024 samples from 0 to 1023 – counting from 0, not 1).

So for a 3.3V reference, and an ADCCODE of 1023 (10-bit ADC), the result will be:

$$
\begin{aligned}
V_{ADC} &= \frac{V_{REF}}{1024} \times ADCCODE \\
&= \frac{3.3}{1024} \times 1023 \\
&= 3.297V
\end{aligned}
$$

Whereas the formula that divides the reference by 1023 will result in a voltage of 3.3V calculated.

Software for displaying ADC voltage

The **`ADC_Voltage_DBGU_port`** program is a modified version of the previous program that converts the ADC codes that are read to voltages and displays them on a terminal emulator. The ADC code for each enabled channel is converted to voltage in the first **for** loop:

```
for (ch_num = 0; ch_num < 4; ch_num++) {
    ADC_voltage[ch_num] = (ADC_VREF / 1023.0) * ADC_data[ch_num];
}
```

The floating point result for each channel is stored in the **ADC_voltage[]** array.

7.1.3 ADC Switch and Keypad Reader

Interfacing switches to the ADC

When there are not enough discrete I/O pins available on the microcontroller for a particular application, it is possible to interface switches to a single ADC channel and thus save I/O pins as shown in **figure 7.3**.

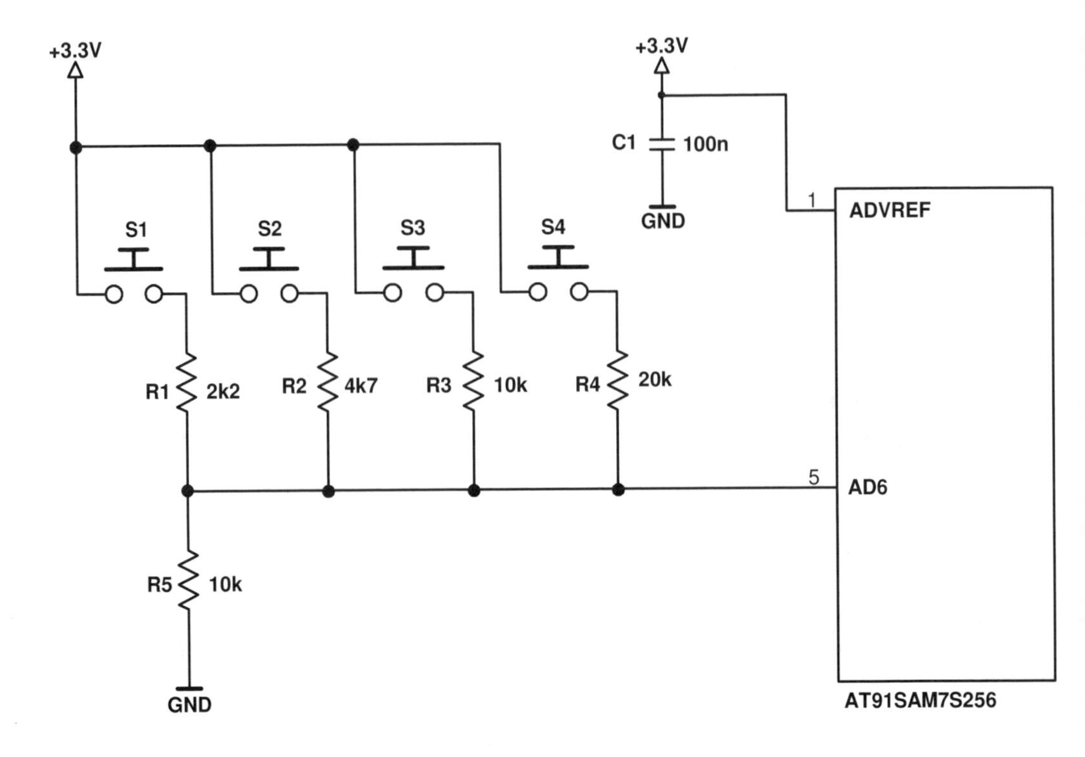

Figure 7.3: Interfacing Switches to an ADC Channel

As each switch is pressed, a voltage divider is effectively connected to the ADC pin. Each switch switches a different voltage divider onto the ADC channel so that the software can distinguish between switches pressed. Many more switches can be interfaced to the ADC channel by adding unique voltage divider values for each switch and reading the voltage in software.

The voltage dividers shown in **figure 7.3** have the following calculated values:

S1 (2k2 / 10k) = 2.7V
S2 (4k7 / 10k) = 2.24V
S3 (10k / 10k) = 1.65V
S4 (20k / 10k) = 1.1V

The following equation was used to calculate these values:

$$V_{DIV} = V_S \times \frac{Rb}{Ra + Rb}$$

Where:
V_{DIV} = Voltage divider voltage on the ADC pin
V_S = Supply voltage (3.3V for **figure 7.3**)
Ra = Top divider resistor (R1 to R4 in **figure 7.3**)
Rb = Bottom divider resistor (R5 in **figure 7.3**)

Software to read the ADC switches
The **`ADC_Switch_DBGU_port_RAM`** program reads ADC channel AD6 to determine if a switch has been pressed. If a switch has been pressed, the program outputs the switch number on the DBGU serial port for display on a terminal emulator.

To determine which switch is pressed, the program converts the ADC code to a voltage and then checks to see if the voltage corresponds to one of the voltages that are expected from one of the voltage dividers connected to the ADC pin. A tolerance was factored into the comparison to compensate for resistor tolerance and noise on the ADC line. E.g. to check for the switch that would put 2.7V on the ADC channel, a voltage of between 2.6V and 2.8V is checked for.

It is not necessary to convert the ADC code to a voltage in this program. A more efficient way would be to calculate what the expected upper and lower ADC codes would be and then check if the ADC code read on the channel falls between these two values.

Interfacing a keypad to the ADC
A keypad can be interfaced to a single ADC channel as shown in **figure 7.4**. The principle of operation is the same as the previous circuit in that each key pressed switches a different voltage divider network onto the ADC pin and therefore a different voltage will be read for each key on the ADC channel.

The figure includes the calculated voltage and the ADC code expected to be read on the ADC channel for each key of the keypad.

Software for the ADC keypad interface
The **`ADC_Keypad_DBGU_port_RAM`** program operates in the same way as the previous program, however it expects closer tolerance voltages (more accurate voltages) to be read. The key pressed, voltage read and ADC code are all sent to the DBGU port for display on a terminal emulator.

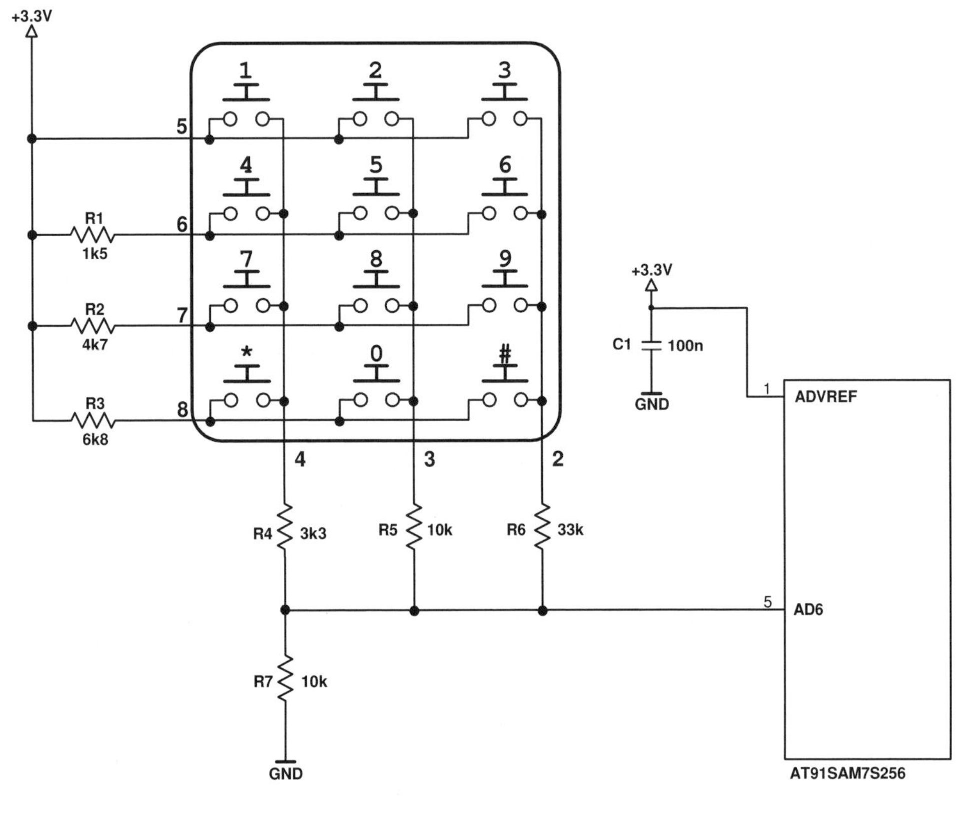

KEY	1	2	3	4	5	6	7	8	9	*	0	#
Voltage	2.48V	1.65V	0.77V	2.23V	1.53V	0.74V	1.83V	1.33V	0.69V	1.64V	1.23V	0.66V
ADC Code	770	512	239	692	475	230	568	413	214	509	382	205

Figure 7.4: Interfacing a Keypad to an ADC Channel

7.1.4 Analogue Signal Conditioning

In order to measure external analogue quantities such as temperature, resistance, voltage, etc. with as high a resolution as possible on the ADC, it is necessary to either amplify or attenuate these external signals when they do not fit into the range that the ADC measures (a voltage between 0 and VREF). E.g. a temperature sensor that outputs a voltage from 0 to 1V over its temperature range will benefit by being amplified so that its maximum output (1V) will appear as 3V on the ADC channel. With a 3.3V reference, this will now use ADC codes from 0 to 931 where if we measured the 1V signal directly it would only use codes 0 to 310.

Attenuating a signal for measuring on the ADC
Resistors used as a voltage divider can be used to attenuate (reduce) a voltage to be measured so that it fits in the range that the ADC can measure e.g. 0V to ADVREF (max 3.3V) on the AT91SAM7S.

Figure 7.5 shows two voltage dividers used to measure voltages that fall outside the range of the ADC inputs on the AT91SAM7S.The divider connected to AD6 was calculated for measuring a voltage between 0V and 12V, but as a precaution can measure a voltage up to 13.2V should the 12V signal be out of tolerance.

The top divider network in the figure scales down a 15V signal to 2.5V. This network can in fact measure voltages up to 19.8V – when 19.8V appears across the divider network, 3.3V will appear on the ADC channel. This compensates for voltages applied to the resistor network that may be out of tolerance. A protection diode, D1 is also included should the input voltage exceed 19.8V. If the input voltage of the resistor divider network exceeds 19.8V + V_F of D1, the diode will conduct and a maximum voltage of 3.3V + V_F will appear on the ADC channel. A schottky diode is shown as it has a lower forward voltage drop (V_F) than a normal silicon rectifier diode.

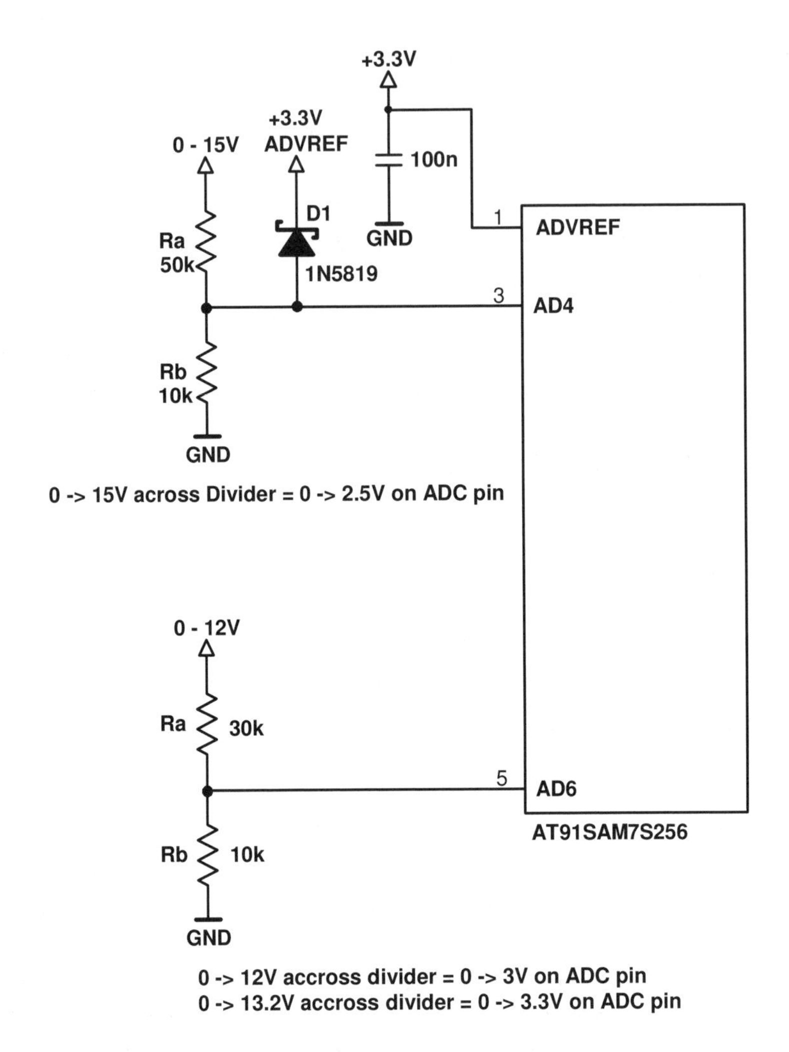

Figure 7.5: Voltage Dividers used to Attenuate a Voltage to be Measured

Software for the voltage divider circuit

The **`ADC_Voltage_div_DBGU_port_RAM`** program reads AD6 and scales the voltage read to reflect the voltage that is being applied across the resistor divider network for 0 to 15V in **figure 7.5**. The resulting voltage is sent to the DBGU port for display on a terminal emulator. The following line of code does the conversion:

```
ADC_voltage = (ADC_VREF / 1023.0) * (float)ADC_data * 6.0;
```

Because the voltage that we are measuring will appear as 2.5V on the ADC channel when 15V is applied across the resistor network, the result of the normal ADC code to voltage conversion that we have already used is multiplied by 6 (15 ÷ 2.5).

To convert the 0 to 12V example of **figure 7.5**, the above line of code must be changed to:

```
ADC_voltage = (ADC_VREF / 1023.0) * (float)ADC_data * 4.0;
```

The scale factor is now 4 (12 ÷ 3).

Amplifying a signal for measuring on the ADC

An op-amp can be used to amplify a small voltage so that it operates the ADC over more of its available range. **Figure 7.6** shows an op-amp configured as a non-inverting amplifier with a gain of 2. I.e. the voltage applied to the input of the op-amp will be amplified to twice its size.

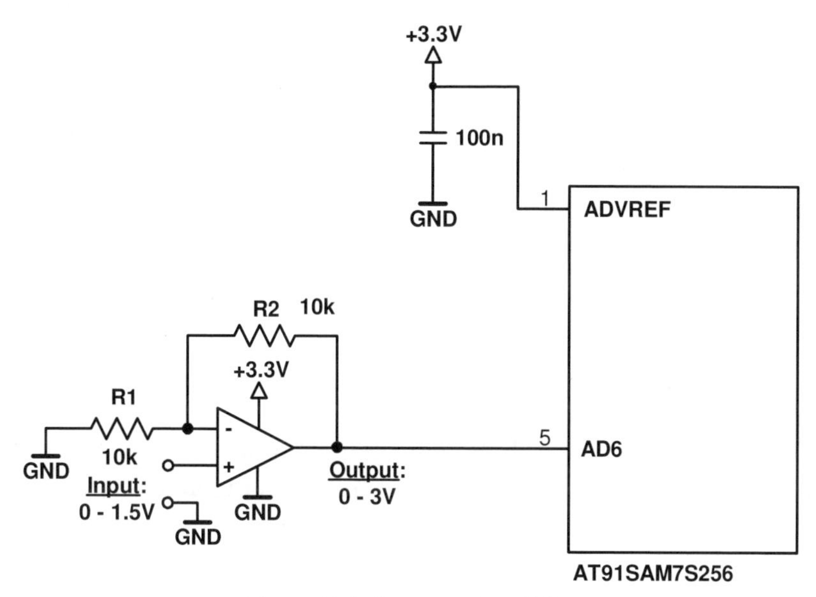

Figure 7.6: Op-amp Amplifier

The formula for calculating the gain of the op-amp is:

$$A_V = 1 + \frac{R1}{R2}$$

For figure 7.6 this results in a gain of 1 + 1 = 2.
Software for the voltage amplifier circuit
The **`ADC_Voltage_div_DBGU_port_RAM`** program can be modified to scale the voltage correctly by dividing the voltage read on the ADC by the gain factor of the op-amp. This is 2 for the example shown.

```
ADC_voltage = (ADC_VREF / 1023.0) * (float)ADC_data / 2.0;
```

7.2 PWM Audio

It is possible to generate an analogue audio signal using the PWM. An audible sine wave can be generated using the circuit of **figure7.7** and the software that follows.

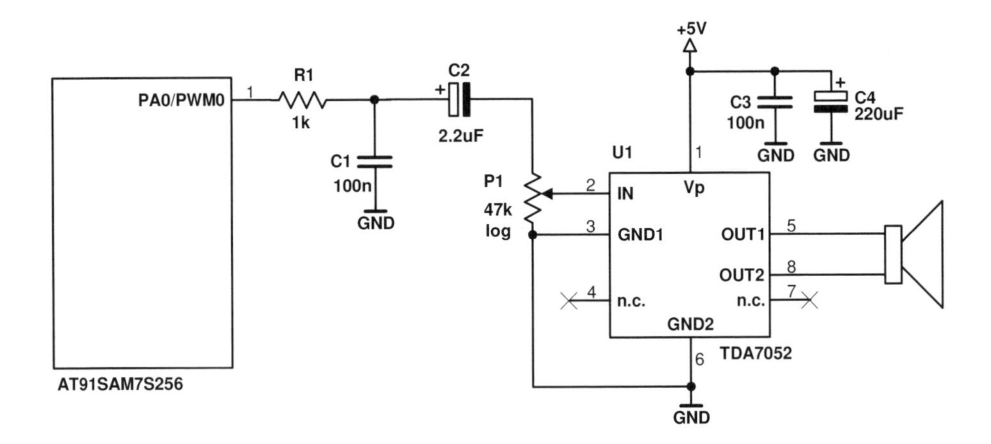

Figure 7.7: PWM Audio Generation and Amplification

In **figure 7.7**, R1 and C1 form a low-pass filter. This allows only low frequencies to pass through to the rest of the circuit which is an audio amplifier. The formula for calculating the cut-off frequency of the low-pass filter is:

$$f_c = \frac{1}{2\pi RC}$$

For the values of R1 and C1 in the figure, the cut-off frequency will be:

$$f_c = \frac{1}{2\pi \times 1k \times 100n}$$
$$= 1592Hz$$

Software for the PWM audio circuit:

The **`PWM_sine_RAM`** program generates a 500Hz sine wave on PWM0. The program uses the PWM interrupt to write each discrete value to the PWM that will make up the sine wave.

After setting up the interrupt controller and enabling pin PA0 as PWM0, PWM0 is set up to be able to generate 256 different pulse widths. This is the equivalent of an 8-bit DAC (Digital to Analogue Converter) as 2^8 (256) different equivalent D.C. voltages can be produced by the PWM. The frequency of the PWM and number of pulse widths that it can produce are set up as follows:

```
PWM0_CMR = 0x00000002;
```

This sets up the clock to the PWM as MCK/4. If the master clock (MCK) is set up to 47.9232MHz, the clock to the PWM will be 11.9808MHz.

```
PWM0_CPRD = 255;
```

Writing 255 to the CPRD (Channel Period Register) will allow 256 different pulse widths to be generated by the PWM (0 to 255 i.e. from 0% to 100% duty cycle) and will result in the PWM output frequency being equal to the PWM input frequency divided by 255. The output frequency of the PWM will then be 11.9808MHz / 255 = 46.983kHz.

Interrupts are enabled so that the pulse width of PWM0 can be changed on each pulse.

In the interrupt service routine, the next value of the sine wave to be produced is set on the PWM output. The value to write to the PWM is obtained from a lookup table. In the program this is the array **`array500[]`** in **`sinewave.c`**.

The sine wave lookup table is generated in a spreadsheet. The following mathematics is used in the spreadsheet to generate the sine wave:

$$y = \sin(\frac{2\pi \times f_{out} \times n}{f_{PWM}})$$

Where:

y = y axis value of the sine wave. This will be 1 at the positive peak of the wave and -1 and the negative peak of the sine wave.

f_{out} = the desired output frequency e.g. 500Hz for the above program

f_{PWM} = PWM frequency

n = sample number from 0 to (**f_{PWM}** ÷ **f_{out}**) - 1

We first need to calculate the number of samples (or voltage levels) that the sine wave will consist of. This is done as follows:

$$n_{SAMPLES} = \frac{f_{out}}{f_{PWM}}$$

For our example in the **`PWM_sine_RAM`** program this will be:

$$n_{SAMPLES} = \frac{46983}{500} \approx 94$$

Now the calculation of **y** must be done 94 times, each time substituting a new value for n from 0 to 93. This can be done in a spreadsheet as follows:

```
=SIN(2*PI()*OUT_FREQ*C9/PWM_FREQ)
=SIN(2*PI()*OUT_FREQ*C10/PWM_FREQ)
=SIN(2*PI()*OUT_FREQ*C11/PWM_FREQ)
•
•
•
=SIN(2*PI()*OUT_FREQ*C102/PWM_FREQ)
```

Where:
OUT_FREQ = 500
PWM_FREQ = 46983
C? = Cell numbers containing the sample numbers 0 to 93

The spreadsheet containing the calculation has been provided for you in the **`code`** directory of the **`PWM_sine_RAM`** project and is called **`sine500.ods`**. It is saved in openoffice.org format. The spreadsheet program called Calc is part of the OpenOffice suit of programs and can be downloaded for free at **www.openoffice.org**.

Figure 7.8 shows the output of the sine wave calculations. Each of the 94 samples that make up the 500Hz sine wave fall in the range of -1 to +1. We need to write values to the PWM in the range of 0 to 255, so the formula is modified as follows to offset the sine wave from 0 and to move it into the desired range:

$$y = 128 \times \sin(\frac{2\pi \times f_{out} \times n}{f_{PWM}}) + 128$$

Because there is a problem with the PWM on some versions of the AT91SAM7S silicon (see chapter 6 - **6.1.1.1 Silicon Errata**), we adjust the calculation to offset the sine wave from 0:

$$y = 126 \times \sin(\frac{2\pi \times f_{out} \times n}{f_{PWM}}) + 129$$

This will give values between 3 and 255.

The **`sine500normalised.ods`** spreadsheet contains the modified calculation:

```
=ROUND(126*SIN(2*PI()*OUT_FREQ*C9/PWM_FREQ)+129)
.
.
.
```

ROUND() rounds off the result of the calculation so that we are left with integers only. The outputs from the calculations can be copied out of the spreadsheet and pasted into a C file and then turned into a C array.

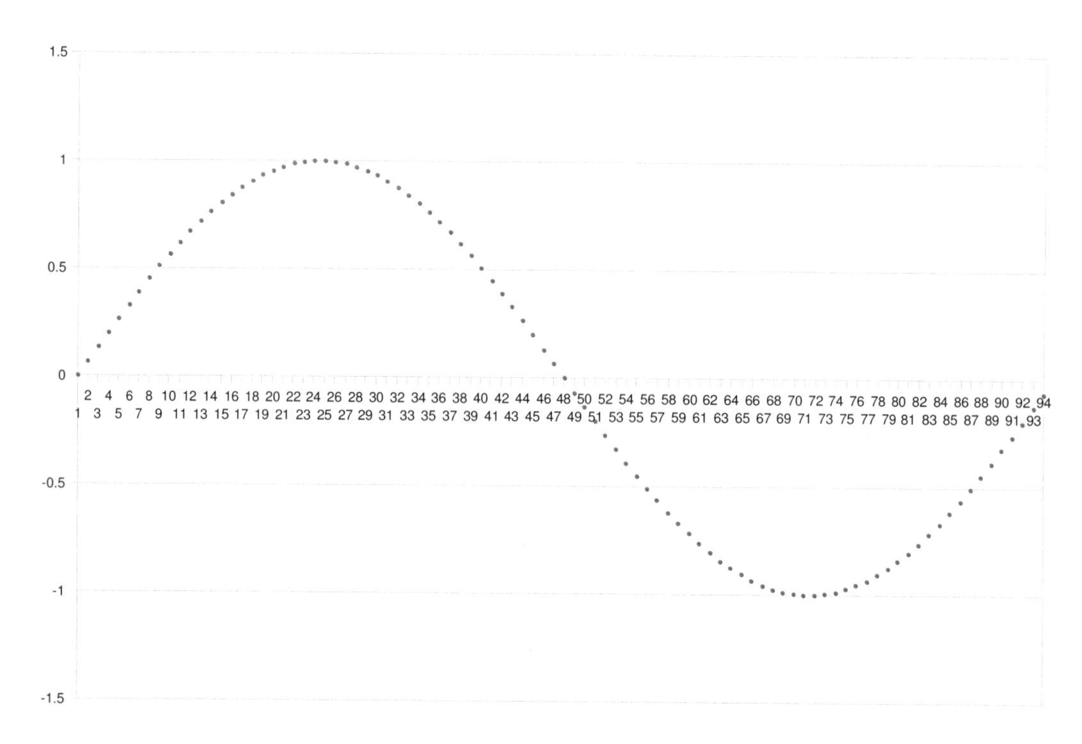

Figure 7.8: Sine Wave Generated by the Spreadsheet

8. Serial Ports

Serial ports transmit data serially i.e. one bit at a time. They allow devices to communicate with each other. As an example, we have been using the DBGU serial port of the AT91SAM7S as an RS-232 port to communicate with the PC.

This chapter looks at four different serial interfaces:

1. **RS-232** – for point to point interfacing of equipment
2. **RS-485** – for serial networking
3. **TWI** – Two Wire Interface: I^2C compatible interface for inter-IC communications
4. **USB** – Universal Serial Bus device port interface

The **SPI** (Serial Peripheral Interface) port is also a serial port, but is discussed in the next chapter on SD cards.

Some serial communication terms that you will come across:

Synchronous – the serial communications are synchronised to a **clock**, one of the lines that connect devices could be a clock line or in some protocols the clock can be part of the data signal e.g. in Manchester Encoding. The two synchronous serial protocols that are covered in this book (TWI and SPI) both have separate clock lines.

Asynchronous – **no clock** line is used, the devices must run at the same speed in order to communicate. Synchronisation does actually occur with the use of a start bit for example.

Baud – for our purposes it means the speed of the serial communications in bits per second. There is a slight difference between baud and bits per second, however they can be taken to mean the same thing in this chapter.

Full duplex – Can send and receive at the same time. e.g. RS-232, telephone – two people can talk simultaneously.

Half duplex – Can send and receive, but only one at a time e.g. RS-485, hand-held radio – push to talk, release to listen.
Simplex – Can send only or receive only, e.g. a radio station transmits only.

UART – Universal Asynchronous Receiver Transmitter. Used to transmit and receive asynchronous serial data. Can be used as a RS-232 or RS-485 port.

USART – Universal Synchronous/Asynchronous Receiver Transmitter. Same as the UART, but can also be used for synchronous transfers if desired.

Table 8.1 compares the serial communication standards discussed in this chapter and the next. The maximum distance that the serial link can operate over is just a guideline as this can change depending on what speed the port is operated at.

Port	Maximum Distance	Maximum Devices	Synchronisation
RS-232	15 – 30m (50-100ft)	2	Asynchronous
RS-485	1200m (4000ft)	32 is standard, but can be up to 256	Asynchronous
TWI	5m (16ft)	40	Synchronous
USB	5m (16ft)	127	Asynchronous
SPI	3m (10ft)	8	Synchronous

Table 8.1: Comparison of some Serial Communication Standards

Figure 8.1 highlights the UART and USART pins on the AT91SAM7S that are available to use as RS-232 or RS-485 ports. USART1 contains full modem line support, however if only the Tx and Rx lines are required, the other pins can be used as general purpose I/O or connected to alternate internal peripherals.

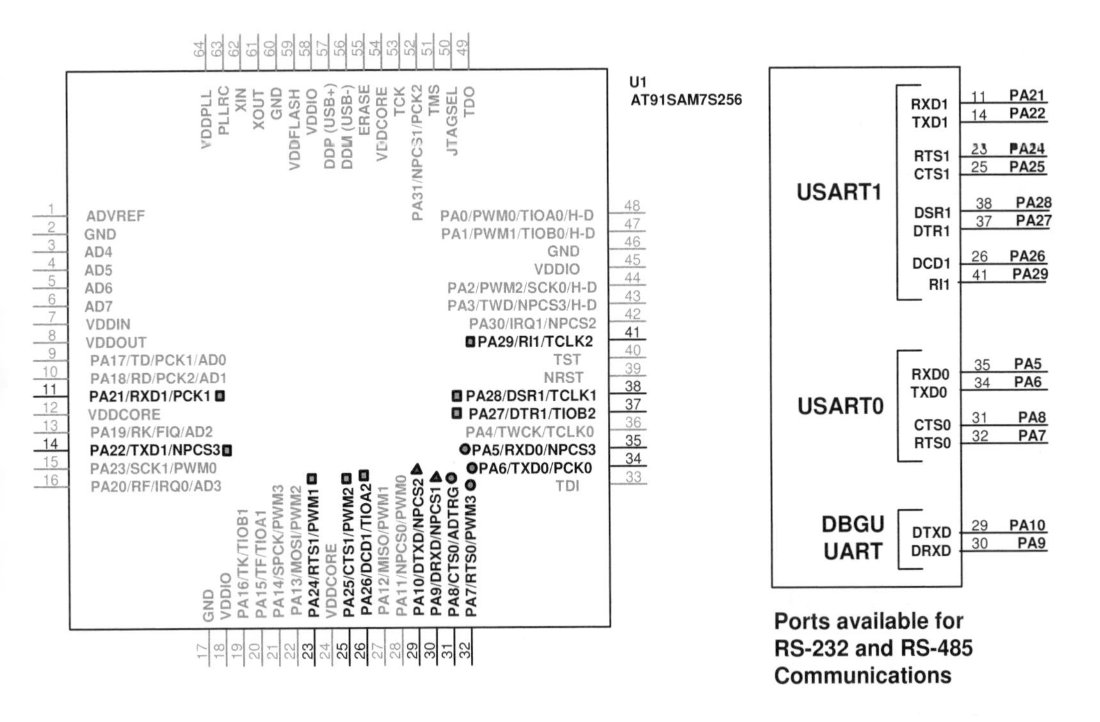

Figure 8.1: The Location of UART and USART pins on the AT91SAM7S

8.1 RS-232

Figure 8.2 shows two RS-232 port configurations. The configuration to the left uses only the transmit and receive pins of USART1 on the AT91SAM7S. The remaining USART pins are free to use as GPIO. The configuration on the right includes the handshaking signals of USART0. Below the schematics is the function of each pin on the DB9 connector when used on an RS-232 interface.

Also refer to **figure 1.8** which shows the DBGU UART interfaced to a RS-232 line driver/receiver IC.

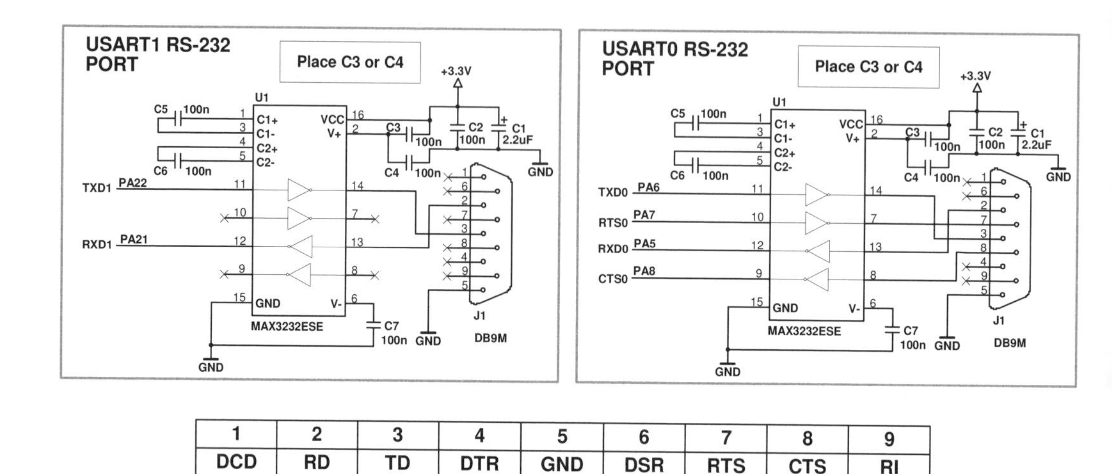

1	2	3	4	5	6	7	8	9
DCD	RD	TD	DTR	GND	DSR	RTS	CTS	RI

Figure 8.2: RS-232 Interfaces

The pins on the DB9 connector have the following functions:

DCD – Data Carrier Detect: signal from modem
RD – Receive Data: serial data is received on this pin
TD – Transmit Data: serial data is transmitted on this pin
DTR – Data Terminal Ready: modem handshaking
GND – Signal ground
DSR – Data Set Ready: modem handshaking
RTS – Request To Send: handshaking signal
CTS – Clear To Send: handshaking signal
RI – Ring Indicator: Modem ring signal

RS-232 line driver/receiver ICs:

The MAX3232E is the RS-232 IC used to interface the UARTs in **figure 8.2**. This IC can operate from a 3V to 5.5V supply. This particular IC that ends with an ‘E’ in its part number provides ESD protection. Some examples of suppliers of the 3232 IC are:
Texas Instruments – e.g. MAX3232E
ST – e.g. ST3232
Some manufacturers specify that the capacitor on the V+ pin of the RS-232 IC must be connected to ground and others specify to Vcc. For this reason it is best to make space for either of these when designing a PCB as you may have to source the IC from an alternate manufacturer. C3 or C4 in **figure 8.2** can then be placed on the PCB depending on the IC used.

Software for the RS-232 serial port
The DBGU port has already been used and software provided for it. Initialising and using UART0 and UART1 is very similar.

8.2 RS-485

RS-485 allows serial communications over a longer distance than RS-232 and also allows more than two devices to be connected to each other – devices can be connected in a network.

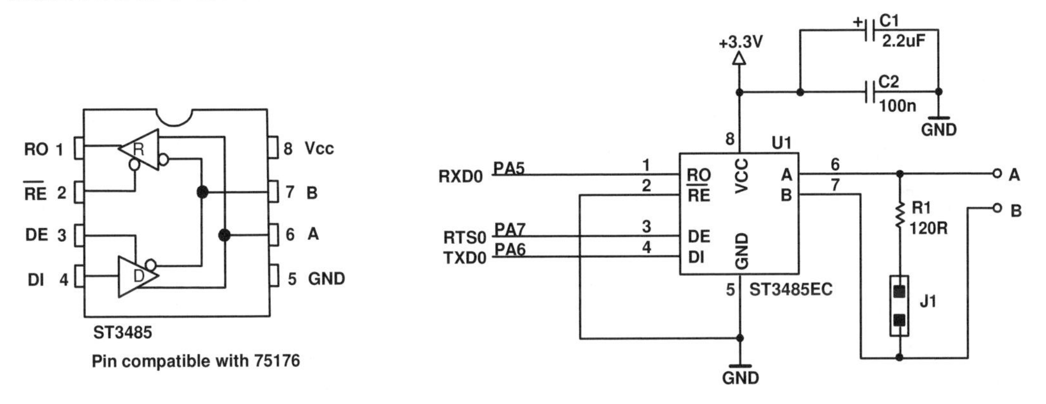

Figure 8.3: RS-485 Interface

Figure 8.3 shows an RS-485 transceiver IC interfaced to USART0 of the AT91SAM7S. The RS-485 transceiver is an ST3485 from ST Microelectronics that operates from a 3.3V supply. The IC is pin compatible with the industry standard 75176 RS-485 transceiver.

Figure 8.4 shows how to network several microcontrollers using an RS-485 network. **Note that the network cable must consist of three lines – A, B and GND**. The ground line must be common between all of the devices on the network.

Depending on the line length, terminating resistors (usually 100 to 120Ω) may need to be placed on the two ends of the network. The longer the network cable, the more likely the terminating resistors will be needed. The circuit in **figure 8.3** provides jumper J1 so that a terminating resistor can conveniently be added at the end of the network by shorting out this jumper. If ten of these circuits were networked together on a long line and it was found that communications were unreliable, only the circuits on either end of the network would have J1 closed to add terminating resistors. The rest of the circuits or "nodes" would leave their terminating resistors open circuit.

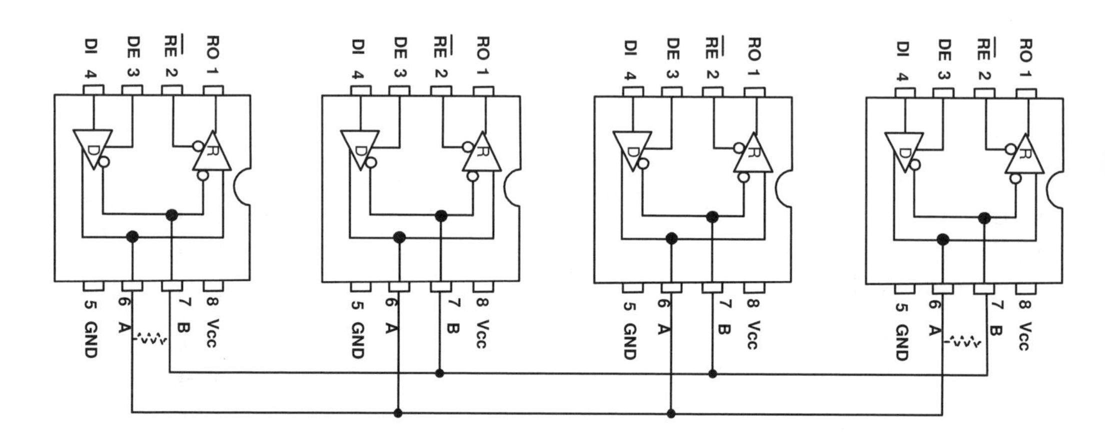

Figure 8.4: A RS-485 Network

RS-485 Network Software

Because the half-duplex RS-485 network shown in **figure 8.4** has both transmitters and receivers on the same lines, a protocol must be devised so that more than one node does not transmit at the same time. If this happens, it would short the transmitters out.

In software, the UARTs connected to the RS-485 chips would all be initialised to the same baud rate as done for RS-232. The pin connected to the DE line of the chip would be pulled low to disable all transmitters.

One of the nodes would act as the master in order to coordinate communications. The master would send a message to each of the slave nodes on the network in turn to see if any of the nodes had anything to transmit. Each slave node would have its own unique address that the master would use to let it know that the data being sent is intended for it.

Each slave node would "listen" on the network to see if the master was transmitting a message containing its ID address. Every node would receive the same message but would reject it if it did not contain its unique ID. If the ID did match, it would be given a certain time to transmit a message. During this time, the master would disable its RS-485 transmitter.

8.3 TWI (I^2C)

The Two Wire Interface (TWI) on the AT91SAM7S is a two wire bus that allows TWI and I^2C (Inter IC) compatible devices to be interfaced to the microcontroller. It allows ICs such as EEPROMs, RTCs, I/O expansion ICs, ADCs and DACs to be interfaced using just two lines and ground.

AT91SAM7S microcontrollers have a built-in TWI controller that can be enabled and connected to two of the microcontroller pins. The two pins operate as TWD (Two-Wire Data) and TWCK (Two-Wire ClocK). These pins are the equivalent of the SDA (Serial Data) and SCK (Serial ClocK) lines on an I^2C bus. The TWI bus can be operated at a maximum speed of **400kHz**.

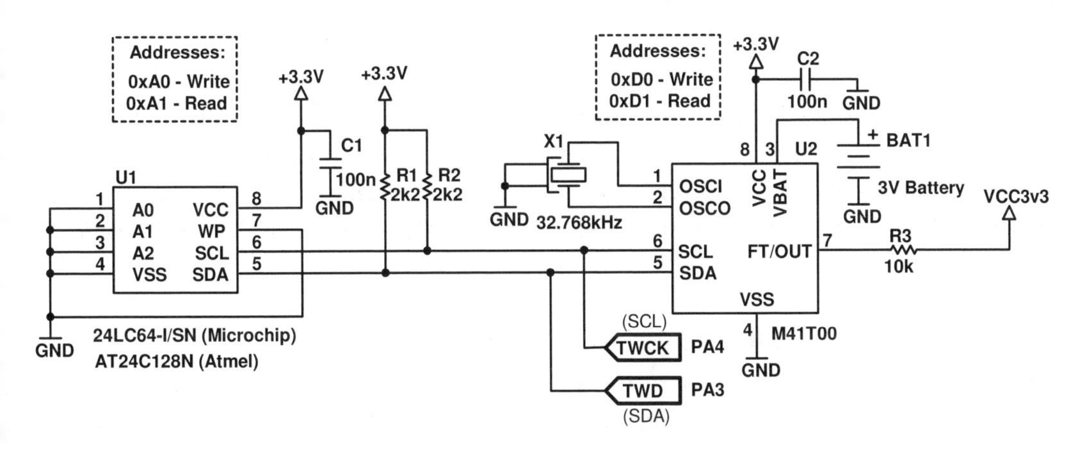

Figure 8.5: Devices Interfaced to the TWI Bus

Figure 8.5 shows two I²C devices connected to the TWI controller of the AT91SAM7S. The device on the left is an EEPROM. Part numbers for two pin compatible EEPROMs from different manufacturers are shown in the figure. The Address pins A0 to A2 allow the address of the EEPROM to be changed so that more than one EEPROM can be attached to the same TWI bus allowing each EEPROM to be uniquely addressed.

A Real Time Clock (RTC) is shown interfaced to the TWI bus to the right in **figure 8.5**. The RTC has a fixed address that is different from the EEPROM allowing it to be individually addressed on the TWI bus. A battery is used to back up the time and date in the RTC when power is switched off.

TWI Software

The **`TWI_EEPROM_RAM`** program reads from and writes to an AT24C128 EEPROM that is interfaced to the TWI bus. The program starts by reading the first ten address locations from the EEPROM and sending them out of the DBGU port for display on a terminal emulator. Normally these values will all be 0xFF on a new EEPROM. The program then writes new values to the first ten locations in the EEPROM. To test that the values were successfully written, switch the power to the microcontroller off and then on and re-run the program. The values previously written to the EEPROM should now be displayed.

The drivers that operate the TWI controller in the AT91SAM7S are located in the **`TWI.c`** and **`TWI.h`** files in the **`Drivers`** sub-directory of the project. The TWI drivers are used by the EEPROM drivers to read/write the EEPROM. The EEPROM drivers are located in **`EEPROM.c`** and **`EEPROM.h`** in the **`Drivers`** sub-directory.

The structure of the `TWI_EEPROM_RAM` program

Figure 8.6 shows the structure of the **`TWI_EEPROM_RAM`** program. **`main()`** calls the low level driver function **`TWIInit()`** directly to initialise the TWI controller. To write a byte to the EEPROM, main calls **`EEPROMWriteByte()`** which in turn calls the low level driver functions **`TWIWrite16Addr()`** and **`TWIPoll()`**. To read a byte from the EEPROM, **`EEPROMReadByte()`** calls the low level driver function **`TWIRead16Addr()`**.

Two additional low level driver functions called **`TWIWrite8Addr()`** and **`TWIRead8Addr()`** have been provided for reading and writing 8-bit data. These functions are not used in the example program, but could be used to access the RTC in **figure 8.5**.

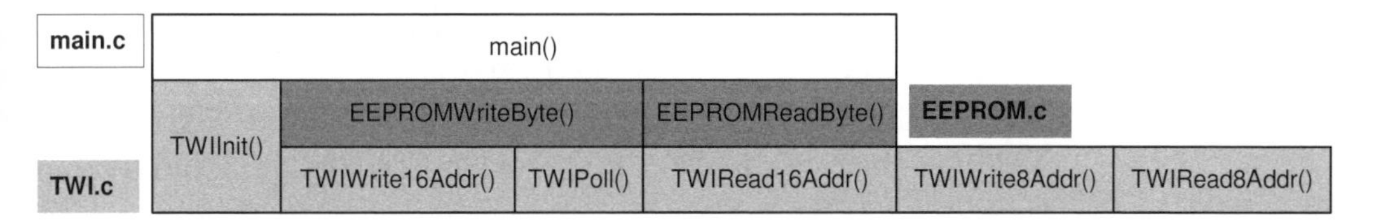

Figure 8.6: Structure of the TWI_EEPROM_RAM Program

`EEPROMReadByte()` is only a "wrapper function" for **`TWIRead16Addr()`** as this is the only function that it calls, calling it with the hardcoded address (0xA0) of the EEPROM (the device address).

The address that is passed to **`EEPROMReadByte()`** and **`EEPROMWriteByte()`** is the internal address of a byte in the EEPROM and has nothing to do with the device address of the EEPROM.

Refer to the well commented driver source code for this project to see how the TWI controller is initialised and used.

8.4 USB

8.4.1 About USB

USB is available on the AT91SAM7S as the USB Device Port (UDP). This allows the AT91SAM7S to be connected as a USB device to a USB host such as a PC.

The USB device port on the AT91SAM7S is a **USB 2.0 full speed (12Mbits per second)** port. The USB 2.0 specification defines three speeds that a USB port can operate at. **Table 8.2** shows these speeds and highlights where the AT91SAM7S USB speed rating fits in.

Speed	M bits per second	Speed Identification Method
Low Speed	1.5Mbits/s	1k5 pull-up resistor on its D- line
Full Speed	12Mbits/s	1k5 pull-up resistor on its D+ line
High Speed	480Mbits/s	Initially uses a 1k5 pull-up on the D+ line to identify itself as a Full Speed device, then performs handshake to negotiate High Speed.

Table 8.2: USB 2.0 Speeds and Speed Identification

A USB connection consists of four lines: D+ and D- (data plus and data minus) for sending and receiving serial data (half duplex – both lines are used to send and both to receive); and +5V and 0V for supplying power to a USB device from the USB host should it be required. When a Full Speed USB device is plugged into a host, it pulls up the USB D+ line with a 1k5 resistor to let the host know that a USB device has been connected and that it is a Full Speed device so that communications can commence at the correct speed. The AT91SAM7S can only operate at Full Speed. The speed identification methods for the other USB 2.0 speeds are shown in the table.

8.4.2 USB Hardware

AT91SAM7S microcontrollers have two dedicated (non-multiplexed) USB pins on the microcontroller, namely DDP (USB D+ line) and DDM (USB D- line).

Interfacing the USB port is very simple as shown in **figure 8.7** as the USB transceiver is on the AT91SAM7S chip and so no external interface chip is needed. The software required to use the USB port on the other hand, is complex.

The figure shows three slightly different configurations. On the left, the 1k5 speed identification resistor can be switched onto the D+ line via a transistor controlled by one of the port pins on the microcontroller. When interfacing to the USB port you can use any of the port pins in your design to control the pull-up resistor. Port pin PA16 is shown in the figure as it is required when using SAM-BA for loading programs to the microcontroller. If SAM-BA is not required, any port pin is fine.

The middle configuration is the same as the first, but allows the device to detect when it is plugged into a host as it will receive a logic high on PA31 – again, any spare port pin can be used to detect this. This is useful if the USB device is externally powered. If it detects that the host is powered off, it can detach the 1k5 pull-up resistor so that current is not fed into the host through this resistor.

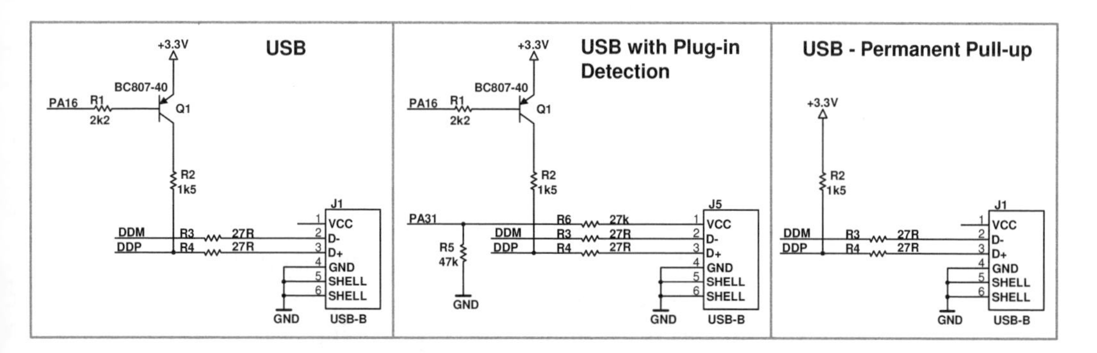

Figure 8.7: Various USB Configurations

The configuration on the right of **figure 8.7** shows a permanent pull-up resistor on the D+ USB line. This configuration is useful to free up a port pin that would normally be used to switch the pull-up resistor into the circuit. It allows SAM-BA to be used, but may confuse the PC software on the host should the embedded system not run the required USB configuration software when it is plugged into the host. This configuration should be avoided if possible.

P-channel FETs can be used in the circuits of **figure 8.7** instead of the PNP transistors. For an alternate arrangement, see the schematic diagram for the AT91SAM7S-EK (found at the end of the "AT91SAM7S-EK Evaluation Board User Guide" – doc6112.pdf from Atmel).

A USB device can be powered in one of two different ways:
Bus powered – the device is powered from the USB port i.e. the host supplies the power.
Self powered – the USB device is powered by a battery or external power supply.

Bus powered devices can fall into one of two categories according to the USB 2.0 specification:
Low power device - consumes less than 100mA
High power device - consumes more than 100mA, but less than 500mA

When a bus powered USB device is initially connected to a host, it must not consume more than 100mA, until the device is configured. Once configured, maximum power can be drawn by the device (up to 500mA).

To use the USB port on AT91SAM7S microcontrollers, an **18.432MHz** crystal must be used in order to generate the correct clock speed for the USB device port running at Full Speed.

8.4.3 More on USB

When a USB device is connected to a USB host, the USB device must perform certain initialisation and configuration routines as the host requests them. These routines are part of the USB specification and this is what makes USB interfacing so much more complex than say an RS-232 port. It is essential to know more about the USB protocol and some terminology before being able to program a USB device. This section provides that information.

Master/slave protocol
A USB device always acts as a slave device on the bus, the host acts as the master. The USB device can not send data to the host unless the host requests it. The host will periodically poll the USB device to send data to it or receive data from it.

Transfer types
USB supports four transfer types namely Control, Bulk, Interrupt and Isochronous.

Control – for configuring the device and controlling it

Bulk – for transfers that have no specific transfer rate requirement e.g. printer, hard drive.

Interrupt – Host polls devices periodically to determine if they have data that needs to be transferred e.g. keyboard, mouse

Isochronous – for real-time synchronous connection e.g. CD audio, speakers, telephones, webcam - guarantees timely data, but not necessarily error free (no error correction)

A USB device must always support the control transfer type. This is necessary for basic communications between the USB host and device. One or more of the other transfer types may be used by the device, depending on what the device does. A USB device can for example support control, bulk and interrupt transfers. It is up to the designer of the USB device to decide which of the transfer types to use.

Endpoints

An endpoint is a send or receive buffer. Endpoints can be set up to transfer data using one of the transfer types (control, bulk, interrupt, isochronous). Every device must have endpoint zero which is bidirectional and used as a control endpoint to establish communications with the host. A device can have up to 30 additional endpoint addresses. Each endpoint has a number (1 to 15) and a direction (IN or OUT).
Direction of the endpoint is defined from the host's perspective:
IN = device to host
OUT = host to device

Endpoints are implemented in hardware on AT91SAM7S microcontrollers. They have four endpoints (0 to 3) and these endpoints can be configured to support the transfer types as shown in **table 8.3**.

Endpoint	Can be configured as	Maximum size
0	Control, bulk or interrupt	8
1	Bulk, isochronous or interrupt	64
2	Bulk, isochronous or interrupt	64
3	Control, bulk or interrupt	64

Table 8.3: Endpoints as Implemented on the AT91SAM7S

USB Descriptors
Descriptors are data structures that contain information about the USB device and its capabilities. The host will request descriptors from the USB device to find out what the USB device is that has been connected to it. The USB specification defines the different types of descriptors that can be used and what information they must contain. Descriptors must be created in software on the USB device and are usually implemented as arrays.

VID/ PID
VID = Vendor ID
PID = Product ID
One of the disadvantages of USB for the hobbyist or small electronics design company is that in order to make a USB device a Vendor ID (VID) number must be obtained and this can only be obtained by subscribing to USB Implementer's Forum (USB-IF) which will cost a few thousand dollars. See **www.usb.org/developers/vendor/**.
The **VID** is a 2 byte number that will be assigned to the vendor (developer i.e. you) after subscribing that will be used when developing USB peripherals. This ID will be sent to the USB host so that the host can identify the device and load the correct device driver. The **PID** is a 2 byte number that is assigned by the vendor/developer and is used to identify the specific USB device. It will also be sent to the host in order for the host to load the correct USB driver. The VID and PID are sent to the host as part of a descriptor.
When doing development on AT91SAM microcontrollers, you can use the Atmel VID and PIDs that have been used in the Atmel example programs.

Pipes
A USB pipe is a logical connection between a USB device's endpoint and the host controller. E.g. a device will always have a control pipe that uses endpoint 0. Other pipes can be established between the device and host.

Enumeration
The process of identifying and configuring a USB device is known as enumeration. When a USB device is plugged into a host, the enumeration process will start. In the enumeration process, the host will assign an address to the USB device, get descriptors from the device and load the appropriate device driver on the host for the program running on the host that will communicate with the USB device.

USB device states

A USB device can be in one of six different states according to the USB specification:

1. **Attached** – device is attached to the USB port
2. **Powered** – the device is powered up
3. **Default** – the device goes into default state after receiving a reset from the host
4. **Address** – after being assigned a unique address by the host
5. **Configured** – device has been configured and can be accessed by the USB driver
6. **Suspend** – the device must enter suspend state if the bus is idle for more than 3ms. In suspend state the device must not draw more than 500μA from the host.

Device classes

A device class is a definition of the functionality of a USB device. There are a number of standard device classes, for example, a USB audio device will belong to the audio device class, a USB RS-232 serial port or modem will belong to the communications device class, a USB keyboard or mouse belong to the HID (Human Interface Device) class. A custom device class can be defined for a custom USB device. The advantage of creating USB devices that use standard device classes is that the USB device developer may not need to write a USB device driver that will run on the host PC. For example, USB flash memory sticks from several manufacturers will all work on the same PC without having to load special drivers. This is because they all belong to the mass storage device class and when detected, the host operating system will load the driver that communicates with mass storage devices.

Steps that occur when a USB device is plugged into a host

This is a simplified description of what happens when a **full speed USB device** is plugged into a host:

1. The USB device is plugged into the host. The device pulls up the USB D+ line with a 1k5 resistor.
2. The host detects the change on the D+ line and knows that a full speed device has been plugged in.
3. The host resets the USB device telling the device to go into its default state (unconfigured, responds only to device address 0 and endpoint 0). It is reset by pulling D+ and D- low for at least 10ms.
4. The host sends a **Get Descriptor** request to the USB device at default address zero.

5. The device sends the **Device Descriptor** packet to the host containing information about the device so that the host can obtain the maximum packet size of endpoint 0 of the device. The host the resets the device again.
6. The host assigns the device a unique 7-bit address on the USB bus by sending a **Set Address** request.
7. The host sends another **Get Descriptor** request and the device responds with the **Device Descriptor** packet containing information about the device including the VID and PID.
8. The host requests various **Configuration Descriptors** from the device to obtain device specific information such as maximum power consumption and number of interfaces.
9. The host requests the **Configuration Descriptor** again and the device responds by sending the **Configuration Descriptor** followed by other descriptors such as the **Interface Descriptor**. Some of the descriptors sent may be class or vendor specific.
10. The host loads the driver needed to operate the USB device. Windows systems use **.inf** files to find the correct driver to load.
11. The USB device is now ready to be operated by the host application. The device can now draw the amount of power from the USB host that was described in the descriptor.

If the USB port cannot supply the amount of power or bandwidth requested by the device during enumeration, the device will not be configured. This might happen if the USB hub that the device is attached to already has other devices attached to it using power and bandwidth.

During enumeration, the device must be ready to respond to any request from the host in any order - not necessarily in the order shown above. The device must also be able to handle a reset from the host at any time.

8.4.4 USB Device Software

The **`USB_Virt_SerialPort_Flash`** program sets up the AT91SAM7S as a USB Virtual COM Port. This means that when it is plugged into the host, it will appear as a serial COM port on the PC. It can easily be connected to by a terminal emulator as it will just appear to be a serial COM port to the terminal emulator program. You can also write your own PC software and then send and receive data via the USB port as if you were communicating with a COM port.

The example program uses the USB CDC (Communications Device Class), however it is not necessary to have any serial port on the USB device as we are simply using the CDC so that we can get data in and out of the AT91SAM7S via the USB port without having to write a USB device driver.

Running the program
This program will run on the AT91SAM7S-EK board. After building the program, load the binary output file **`USB_Virt_SerialPort_Flash.bin`** to the AT91SAM7S embedded system using SAM-BA. The first time that the device is plugged into the USB port, Windows will want to install a device driver for it. Direct it to the **`Win_Driver`** subdirectory of this project. This directory contains the INF file **`6119.inf`** that contains the driver information necessary to set up the USB device.

After the device has been successfully installed, the COM port number that was assigned to it can be found as follows:
Right click on "**My Computer**", click **Properties**, click the **Hardware** tab, click the **Device Manager** button. If the device was successfully installed, it will appear under the **Ports (COM & LPT)** item in the **Device Manager** window. Click the '+' sign to expand the item and view the ports installed. It should display **AT91 USB to Serial Converter (COM15)** where COM15 could be any port number that was assigned to the USB device.
Now that you know the COM port number, it is possible to connect it to a terminal emulator. When connected, pressing any key on the keyboard will initially send a message from the AT91SAM7S to the terminal emulator. Every time a key is pressed, the USB device will respond by sending a message back to the terminal emulator stating which key was pressed.

You now have basic USB communications operating and can modify the program as required to suit your application.

Changing parameters and settings
The example program was adapted from the **usb-device-cdc-serial-project-at91sam7s-ek** Atmel example program and so borrows its VID, PID and INF file.
The vendor ID is: **0x03EB** (Atmel)
The product ID is: **0x6119**

These parameters can be changed in the **devDescriptor[]** array in the **USB_CDC.c** file. They are well commented in the source file so you should have no problem finding them. In the array, they appear with the least significant byte first.

Another setting that you may want to change is the port pin used to switch the 1k5 pull-up resistor onto the USB D+ line. This can be found in the **USBInit()** function in the **USB_CDC.c** file. It is currently set to operate port pin PA16.

How the program works

<u>The structure of the program:</u> **USB_CDC.c** and **USB_CDC.h** in the project's Drivers sub-directory contain everything pertaining to USB used in the project. The program does not use interrupts, but polls for USB events in the **while(1)** loop.

At the start of the program, **USBInit()** is called to initialise USB hardware. It enables pin PA16 as an output pin in order to control switching of the 1k5 pull-up resistor onto the D+ line. The clocks used by the USB port are then enabled. The pull up resistor is switched onto the D+ line.

USBConfigure() is called next inside a **while** loop so that it will continually be called until USB enumeration has succeeded. **USBEnumerate()** is called from within **USBConfigure()** to enumerate the USB device.

USBEnumerate() takes care of control messages sent from the host. It will send the descriptors to the host when asked.

Only after enumeration has taken place will the **while(1)** loop be entered. **USBRead()** is called to check for and receive data coming from the USB application running on the host e.g. a terminal emulator program.

USBTxMsg() is called to transmit a message to the host USB application should data be received. This function will send C strings as it checks for the null terminator in the string to determine its length. **USBWrite()** is called from **USBTxMsg()** to send the data.

8.4.5 Atmel USB Example Programs

Atmel provides AT91SAM7S USB example programs for implementing USB enumeration, a USB to serial RS-232 port converter, USB keyboard, USB mouse and USB mass storage device. These example programs will run on the AT91SAM7S-EK board. The projects are listed below with additional details and a reference to the documentation available for them. You can use these projects to start developing your own USB peripheral devices.

- USB ENUMERATION-

Document: **AT91 USB Framework** (doc6263.pdf)
Software: **`usb-device-core-project-at91sam7s-ek`**
Performs USB enumeration only.
Vendor ID: 03EBh
Product ID: 0001h

- USB TO SERIAL CONVERTER -

Document: **AT91 USB CDC Driver Implementation** (doc6269.pdf)
Software: **`usb-device-cdc-serial-project-at91sam7s-ek`**
This project implements a USB to Serial converter on the AT91SAM7S using the USB device port and USART0. It demonstrates the use of the Communication Device Class (CDC).

Vendor ID: 03EBh
Product ID: 6119h
Driver file: 6119.inf

Resources used:
Endpoint 0: control endpoint
Endpoint 1: Bulk OUT
Endpoint 2: Bulk IN
Endpoint 3: Interrupt IN
USART0

- USB HID – MOUSE AND KEYBOARD -

Document: **AT91 USB HID Driver Implementation** (doc6273.pdf)
Implements HID (Human Interface Device) mouse and keyboard projects.

HID mouse
Software: **`usb-device-hid-mouse-project-at91sam7s-ek`**
Moves the mouse cursor using the buttons on the evaluation board
Vendor ID: 03EBh
Product ID: 6200h

HID Keyboard
Software: **`usb-device-hid-keyboard-project-at91sam7s-ek`**
Sets the AT91SAM7S-EK board up as a HID Keyboard Device.
Vendor ID: 03EBh
Product ID: 6127h

Resources used:
Uses the following push buttons on the board:
PA19 types the character 'a'
PA20 toggles the Num Lock LED on the keyboard
PA15 types the character '9'
PA14 is configured as a Shift key

LED DS3 (on PA2) is configured as the Num Lock LED.

- USB MASS STORAGE DEVICE -
Document: **AT91 USB Mass Storage Device Driver Implementation** (doc6283.pdf)
Software: **`usb-device-massstorage-project-at91sam7s-ek`**
Implements a Mass Storage device in internal Flash memory of the AT91SAM7 microcontroller. When the board is plugged into a PC, it will be recognised and configured as a mass storage device.
Vendor ID: 03EBh
Product ID: 6129h

Additional Atmel USB documentation:
AT91SAM7S64 USB Certification (doc6213.pdf) contains information on the USB certification process of a USB V2.0 Full Speed device.

9. SD Memory Cards

Secure Digital (SD) memory cards allow you to add hundreds of megabytes or even gigabytes of data storage to your embedded system at relatively low cost. This can be useful for tasks such as data logging. SD cards are widely available and used in electronic equipment such as digital cameras and portable digital music players.

An SD card can be interfaced to a microcontroller's **SPI (Serial Peripheral Interface)** port. Data can be written to the SD card in raw format - i.e. byte for byte starting from address 0, but it is more useful to implement the FAT file system so that the card can be read and written by both an embedded system and a PC.

This chapter provides information on SD cards, the SPI port and the FAT file system so that you will be able to interface an SD card to your embedded system and write to it using the FAT file system.

9.1 SD Memory Card Basics

9.1.1 Introduction and Card Sizes

SD memory cards can be interfaced to a system in one of two ways – either using the proprietary SD card protocol that uses a 4-bit data bus (there is also a 1-bit SD transfer mode), or SPI mode that clocks data in and out of the device on two pins (one for shifting serial data into the device and one for shifting serial data out of the device). To get the specification document for the proprietary SD card protocol requires you to join the **SD Card Association** at a cost of a few thousand dollars per year. This makes it unaffordable for hobbyists or companies doing small runs of a product that incorporate SD cards. A simplified specification is available to download for free from **www.sdcard.org** – the simplified physical layer specification version 2.00. To download the simplified specification document you will need to accept the conditions which the SD Card Association has set on this page: **www.sdcard.org/developers/tech/sdcard/pls/**

SD card specification versions

Any SD card that you buy may comply with the new version 2.00 specification or with one of the older version 1.x specifications e.g. version 1.10. The version of card interfaced to a microcontroller can be detected by reading registers in the card and the slight differences between the different version cards can be handled in software – mainly during initialising the cards.

SPI communications

The SPI protocol is open and since many microcontrollers have a SPI port, we will be interfacing our SD cards to the SPI port. The only disadvantage of using SPI compared to the native SD protocol is that it will be slower.

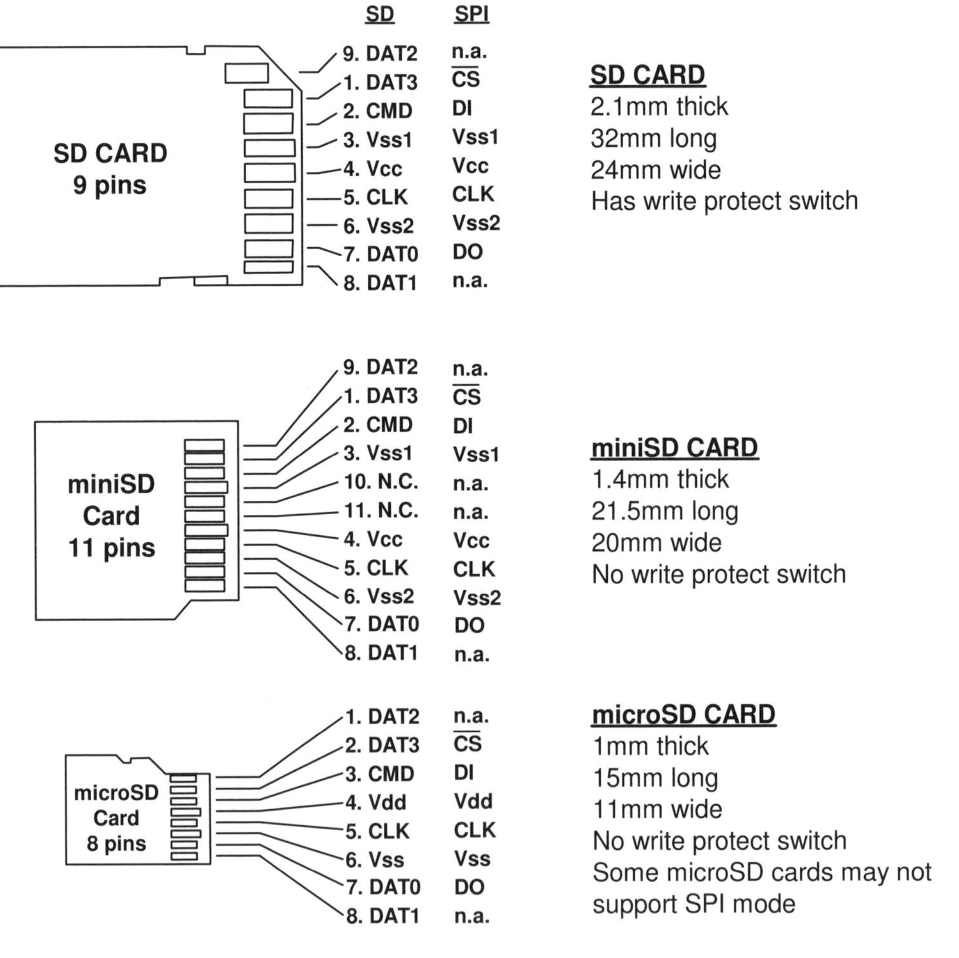

Figure 9.1: SD Card Types and Pin Numbers

SD card physical sizes and forms
Since the original SD card was produced, two additional physically smaller SD cards have become available, namely the miniSD and microSD cards. These cards will operate from the same software as the original SD card, however the microSD card is not required to support SPI mode so it may be wise to avoid buying these cards for an embedded system unless you are sure that this mode is supported on the particular card that you are going to buy. Adapters are available to convert the smaller SD cards to fit into a normal SD card socket.

Figure 9.1 shows the various SD card forms, dimensions and pin numbering. The pin functions to the left show the names of the pins when used in the proprietary SD protocol and the pin functions to the right show the names when used in SPI mode.

Internally, an SD memory card consists of an intelligent controller interfaced to Flash memory.

9.1.2 Supply Voltage and Current

SD memory cards operate from supply voltages in the range of **2.7 to 3.6V** so can operate directly from a microcontroller's 3.3V supply. The cards can draw around **75mA to 100mA** of current, so it is a good idea to design a system that can supply more than this amount of current. The current consumption mentioned here is the maximum value that you will most likely encounter, however, one of the SD cards internal registers (CSD register) contains the maximum read and write current consumption and this register can contain values from **1mA up to 200mA** (version 1.x cards only). An example of SD card current consumption from SanDisk (**www.sandisk.com**) is as follows:
100mA maximum at 25MHz for read and write operations.
200mA maximum at 50MHz for read and write operations.

Provision has been made to support dual voltage SD card in the future (starting with version 2.00 cards). These cards will be able to operate from the "High Voltage" (2.7V to 3.6V) and be able to operate from a lower voltage as well.

9.1.3 Memory Capacity

SD memory cards are available in two memory capacity categories:

1. Standard capacity – up to **2GB**
2. High capacity (SDHC) – up to **32GB**

A host that is operating from the version 1.00 SD specification will fail to initialise a high capacity SD card which was first defined in the version 2.00 SD specification.

A third type of SD card is soon to become available – the SDXC card (extended capacity) that will support capacities up to **2TB**. Access speeds of the new cards will be faster than the previous two SD card types.

9.1.4 Clock Speed

The SD version 1.00 specification states that a full speed SD card will operate in the range of **0 to 25MHz**. The SD card 2.00 specification states that in default mode, cards will operate from **0 to 25MHz** and in high speed mode will operate from **0 to 50MHz**.

9.1.5 Write protection

Operation of the write protect tab on the standard SD card is not implemented in the card, but in host software (e.g. in the embedded system). The SD card socket must have a switch that is triggered by the write protect tab. Write protection of particular sections of memory in the card can also be supported by SD cards. These areas can be specified as write protected by writing to the card internal registers in software.

9.2 SPI Port

9.2.1 SPI Port Operation

Many microcontrollers today have a SPI (Serial Peripheral Interface) port, including AT91SAM7S microcontrollers. The SPI port operates using synchronous serial transfer of data (i.e. transfers are synchronised to a clock) and is full duplex. It consists of four lines:

MOSI – Master Out Slave In: (serial data line) shifts data from the microcontroller (master) into the SD card or other SPI slave device.
MISO – Master In Slave Out: (serial data line) shifts data from the SD card or other SPI slave device into the microcontroller (master).
SPCK – (or SCLK) Serial Clock: (clock line) clock pulse from the master for clocking data bits in and out of the SPI port.
NSS – (or SS) Slave Select: (chip select line) Chip select line of the SPI slave device e.g. to select the SD card.

Several slave devices can be connected to the SPI bus and are selected by the master using an individual chip select line from the master to each slave device.

Data is clocked in and out at the same time i.e. when a byte is clocked into the slave from the master, at the same time a byte from the slave is clocked into the master, bit for bit. If either the master or slave does not need the byte that is clocked into it, it can be ignored – e.g. if a byte is sent to the slave device, the master may not need the byte that is clocked out of the slave at the same time, so it will be ignored by the master. To read a byte from the SD card, 0xFF is written to the card, thus clocking a byte out of the card at the same time.

Data is shifted in and out of the SPI port most significant bit first of each byte.

9.2.2 AT91SAM7S SPI Port and SD Card Interfacing

The AT91SAM7S SPI port can be configured as a master or a slave. This is useful for communication between microcontrollers. To interface to an SD card, the microcontroller's SPI port is set to master mode.

Figure 9.2 shows how to interface an SD card (socket) to an AT91SAM7S microcontroller.

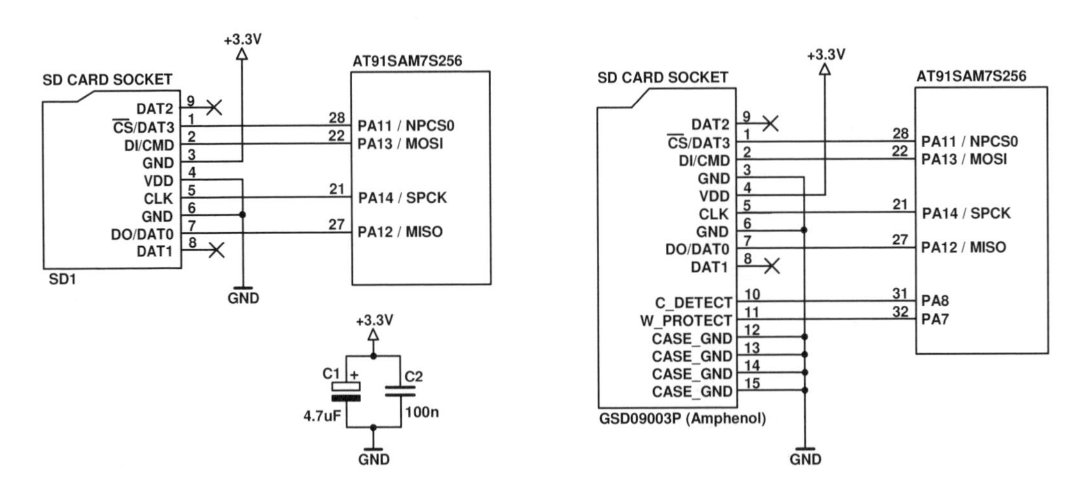

Figure 9.2: Interfacing an SD Card to an AT91SAM7S Microcontroller

The circuit on the left in **figure 9.2** shows the minimum requirements to interface an SD card to the AT91SAM7S. The circuit on the right includes interfacing the SD card socket's card detection and write protect pins. On the microcontroller, these pins must be set as GPIO inputs and can be read to see if a card has been inserted into the socket and if the write protect switch is on or off.

Normally **10k pull-up resistors** would be required on the SPI bus lines as well as the card detect and write protect pins, but this design relies on the microcontrollers internal pull-up resistors being switched on.

The microcontroller pin connected to the chip select line of the SD card as shown in the figure is a dedicated chip select line from the microcontrollers SPI port. The microcontroller has in fact four such chip select lines (NPCS0 to NPCS3) that can be used to interface up to four SPI devices to the same data and clock lines. It is not necessary to use these specific pins as other microcontroller pins can be set up as GPIO output pins and manually controlled to select the SD card. The software examples that are presented in this chapter set up the chip select line as a GPIO output as this gives the programmer more control over this pin. To implement this simply do not assign the dedicated chip select pin(s) to the SPI port. The SPI data and clock lines will need to be configured as SPI pins.

When laying out the SPI signal tracks on a PCB it is a good idea to keep the length of the tracks to a minimum to ensure reliability at maximum speed.

9.2.3 SD Card Sockets

There are a number of different SD card sockets available from different manufacturers. Some will lock the card into the socket when pushed in and the card must then be pushed again to unlock and remove it. Others are the simple push in, pull out type such as the GSD09003P from Amphenol shown in **figure9.3**. To the best of my knowledge, SD card sockets are only available as surface mount devices, although some companies offer SD card sockets soldered to a break out board allowing easy soldering of wires to the socket.

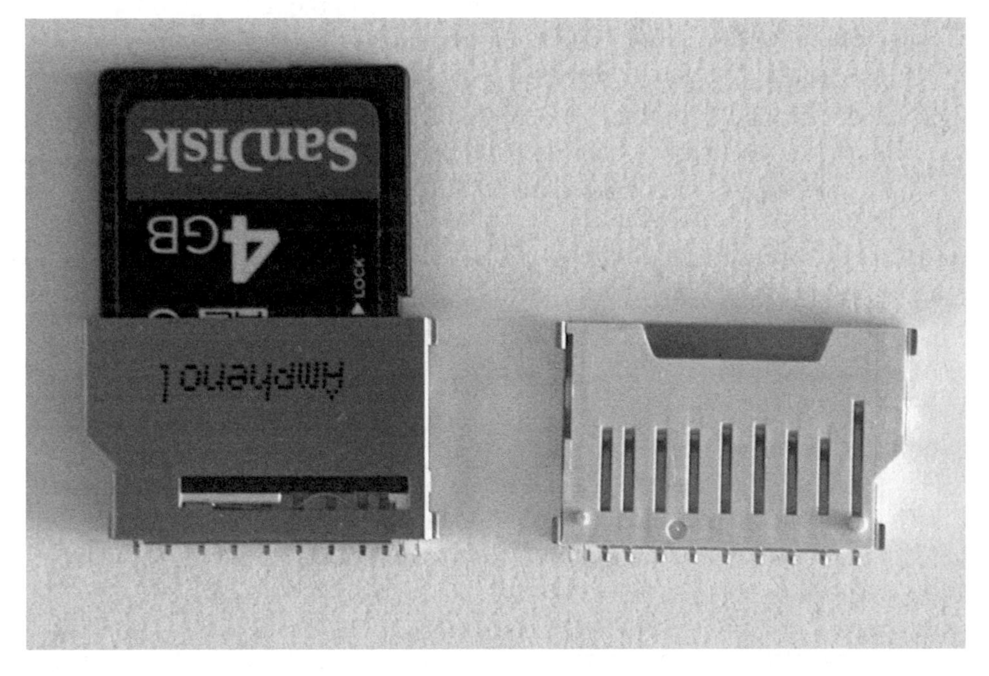

Figure 9.3: SD Card Socket from Amphenol – Top View (left) and Bottom View (right)

9.3 SD Memory Card Operation in SPI Mode

9.3.1 SD Card Protocol in SPI Mode

At power-up, the SD card starts up in SD mode and must be put into SPI mode. Commands or data blocks can be sent to the SD card, a response or data block can be received from the SD card. In **SPI mode**, SD card communications have the following features:

- Byte oriented operation – every command or data block consists of bytes.
- Commands and data blocks are byte aligned to the /CS signal.
- Data is shifted in or out most significant byte (MSB) first
- Communications are controlled by the host (microcontroller).
- Standard capacity memory card data blocks can be as small as one byte.
- High capacity memory card data blocks are fixed at 512 bytes.
- CRC checking is not enabled by default in SPI mode.

A command can be sent to the SD card from the host e.g. to request the contents of one of the cards internal registers. A response is received back from the SD card in response to a command. Data can be sent to the SD card to be stored on the SD card or read from the SD card.

SD command format

All SD commands in SPI mode consist of six bytes as shown in **figure 9.4** and sent from the host (microcontroller) to the SD card. If any of the argument fields in the command packet are not required for a specific command, they must be set to zero.

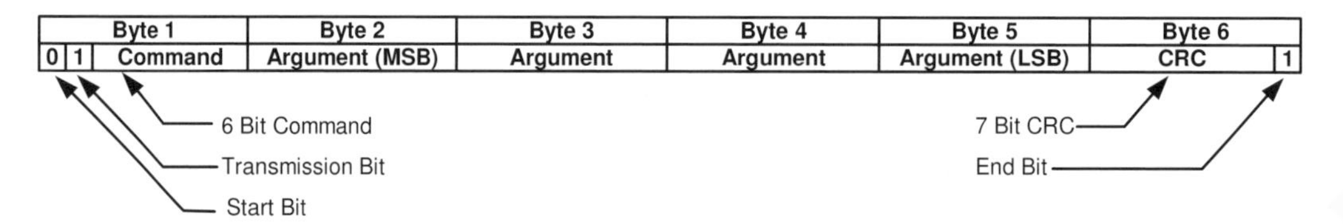

Figure 9.4: SD Card Command Format in SPI Mode

Commands are named according to their command value. e.g. CMD0 is the command 000000b in the 6 bit command field, CMD8 is 001000b. **Table 9.1** lists some of the most important commands that are valid in SPI mode. For other commands, see the SD card simplified physical layer specification version 2.0 document.

Table 9.2 lists some of the application specific commands available in SPI mode. To send an application specific command, first send CMD55. The next command that is sent to the card will then be interpreted as an application specific command.

Explanations of the registers and tokens mentioned in the tables will follow.

Command / Response	**Argument**	**Abbreviation**	**Description**
CMD0 / R1	0	GO_IDLE_STATE	Resets card
CMD8 / R7	[31:12] = 0 [11:8] = supply voltage (VHS) [7:0] = check pattern	SEND_IF_COND	Sends supply voltage that the host operates at. Gets the SD card voltage operating range.
CMD9 / R1	0	SEND_CSD	Gets the contents of the SD card's CSD (Card Specific Data) register
CMD10 / R1	0	SEND_CID	Gets the contents of the SD card's CID (Card Identification) register
CMD12 / R1b	0	STOP_ TRANSMISSION	Stops the card from transmitting during a Multiple Block Read Operation
CMD13 / R2	0	SEND_STATUS	Asks the SD card to send it's status register
CMD16 / R1	Block length in bytes	SET_BLOCKLEN	Sets the block length of standard capacity SD cards for read and write operations
CMD17 / R1	[31:0] data address in bytes for standard capacity cards, or block address for high capacity memory cards (512 byte block address	READ_SINGLE_ BLOCK	Reads a block of data from the SD card at the specified address.
CMD18 / R1	As above	READ_MULTIPLE_ BLOCK	Continuously transfers data blocks from the SD card until the host issues a STOP_TRANSMISSION command
CMD24 / R1	As above	WRITE_BLOCK	Writes a block of data to the SD card at the specified address

CMD25 / R1	As above	WRITE_MULTIPLE _BLOCK	Continuously writes blocks of data until a 'Stop Tran' token is sent instead of 'Start Block'
CMD55 / R1	0	APP_CMD	Tells the SD card that the next command will be an application specific command (ACMD) – see **table 9.2**
CMD58 / R3	0	READ_OCR	Reads the OCR (Operations Conditions Register) of the SD card.

Table 9.1: Some of the SD Card Commands Available in SPI Mode

Command / Response	Argument	Abbreviation	Description
ACMD13 / R2	0	SD_STATUS	Read the contents of the SSR (SD Status Register)
ACMD41 / R1	[31] = 0 [30] = HCS [29:0] = 0	SD_SEND_OP _COND	Sends HCS (Host Capacity Support) bit to the SD card. Starts the SD card's initialisation process. HCS must be set to 0 if the host does not support high capacity cards, or to 1 if it supports high capacity cards
ACMD51 / R1	0	SEND_SCR	Gets the contents of the SD card's SCR (SD Configuration Register)

Table 9.2: Some of the SD Card Application Specific Commands Available in SPI Mode

SD response format

Each command that is sent to the SD card will receive a response back from the SD card as shown in **table 9.1** and **table 9.2** under the Command/Response heading. Some of these responses will be followed by a token and then a block of data. Details of the tokens and data blocks are given under the SD card registers section and the reading and writing the SD card section that follow. **Figure 9.5** shows the details of the responses that can be received back from the SD card.

R3 response / OCR register

The R3 response contains the contents of the OCR register. Bit 30 of the OCR is the CCS (Card Capacity Status) bit. This bit will be set if the SD card is a high capacity memory card and is only valid if bit 31 (card power up status) is also set. Bits 15 to 23 contain the voltage profile of the card. Refer to the simplified SD card specification for details.

R7 response / CMD8

After sending a CMD8 command, an R7 response is returned by the SD card. The 'check pattern' field in the R7 response must contain the same check pattern that was sent in the argument field of the CMD8. The SD card documentation suggests that this check pattern should be set to 10101010b (0xAA). For our purposes, the 'voltage accepted' field should always be 0001b which means that the card operates from 2.7V to 3.6V.

Sending a command to the SD card, polling for a response

After a command has been sent to the SD card, the microcontroller (host) must read back bytes from the SD card i.e. poll it until it gets an R1 response. Every response starts with and R1 response, so the microcontroller must poll for the R1 response and then depending on the command that was sent, may have to read the other bytes of the response that will directly follow the R1 response.

To poll for a response, the microcontroller writes a byte with the value of 0xFF and then reads the response after each byte is sent. It is necessary to write 0xFF to the SD card as data must be clocked out of the SPI port in order to clock data into the SPI port. The microcontroller will read back 0xFF from the SPI port until the R1 response is received. Because bit 7 of the R1 response is always zero, the microcontroller can poll for any byte that is not equal to 0xFF. This will then be the R1 response.

The R1 response is expected back in within 8 to 10 bytes of polling. If it is not received, it is considered a timeout.

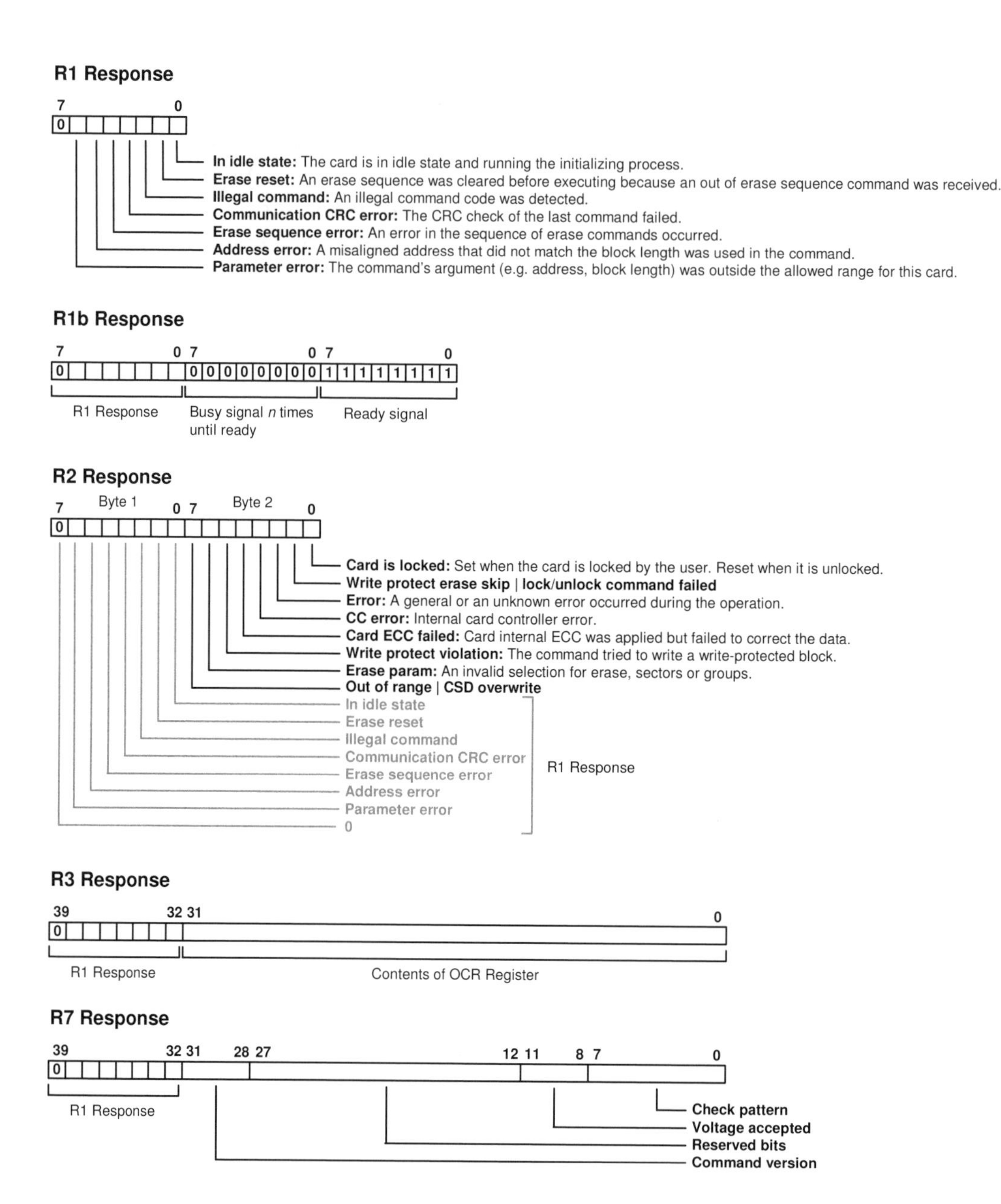

Figure 9.5: SD Card Response Format

9.3.2 SD Card Registers

Table 9.3 shows the internal SD card registers that are available in SPI mode. Refer to the simplified SD card specification version 2.0 for details of the contents of these registers – there is a lot of information stored in these registers and it is pointless reproducing it in this book. The CSD register has two different versions. The CSD structure inside this register is common between different version CSD registers, so software can be used to determine which version of the register has been read. The fields in this register can then be interpreted correctly.

Register	Width	Description
CID – Card Identification	128	Unique identification number which includes manufacturer ID, product name, manufacture date and more
DSR – Driver Stage Register	16	To configure the card's output drivers (Optional)
CSD – Card Specific Data	128	Information regarding access to the card's contents
SCR – SD Configuration Register	64	SD memory card's special features capabilities information
OCR – Operation Conditions Register	32	Contains operating voltage ranges and status bits
SSR – SD Status Register	512	Information about the card proprietary features such as speed class, bus width and more
CSR – Card Status Register	32	Card status information containing error conditions and more

Table 9.3: SD Card Registers Available in SPI Mode

Reading the SD card registers

To read the SD card registers, the following commands are written and responses read back:

CID – command: CMD10, response: R1

DSR – simplified specification 2.00 says that the DSR is available in SPI mode, but the command (CMD4) for accessing it is not available in SPI mode.

CSD – command: CMD9, response: R1

SCR – command: ACMD51, response, R1

OCR – command: CMD58, response: R3

SSR – command: ACMD13, response: R2

CSR – command: CMD13, response: R2

Only 16 bits can be read from the CSR when in SPI mode in the form of the R2 response.

A description of what must take place to read each register is shown below. The timing on the SPI bus included after the descriptions. SPI timing information is not included in the simplified SD card specification and had to be obtained from SD card manufacturer's datasheets. More information on the types of tokens used in the SD card protocol is included in the section on reading and writing SD cards.

Reading the CID and CSD registers

Send CMD9 (CSD register) or CMD10 (CID register) with argument field set to zero, get R1 response, wait for response token (0xFE) – keep reading bytes until 0xFE read or timeout occurs, read 16 bytes (the register contents), read the two CRC bytes

Reading the OCR register

Send CMD58 with the argument field set to zero, read an R3 response that consists of an R1 response and the contents of the OCR register in the next four bytes.

Reading the SCR register

Send a CMD55 to notify the SD card that the next command will be an ACMD, read back an R1 response, send the ACMD51 and read back an R1 response. Poll for the response token (0xFE), the next 8 bytes will be the contents of the 64 bit SCR register. Two CRC bytes must now be read.

Reading the SSR register

Send a CMD55 to notify the SD card that the next command will be an ACMD, read back an R1 response, send an ACMD 13 command, read back an R2 response that consists of an R1 response and an extra byte. Poll for the response token (0xFE), the next 64 bytes after the token will be the 512 byte SSR register contents. Read two CRC bytes.

Reading the CSR register

Send a CMD13 and read back an R2 response which consists of an R1 response followed by a second byte.

Timing when reading the SD card registers

Figure 9.6 shows the timing used when getting the SD card registers. Extra clock cycles must be added to the register read transactions as shown by writing bytes with the value 0xFF to the SPI port. This information is not available in the simplified SD card specification but comes from the MMC (Multi Media Card) documentation and from datasheets of SD card manufacturers.

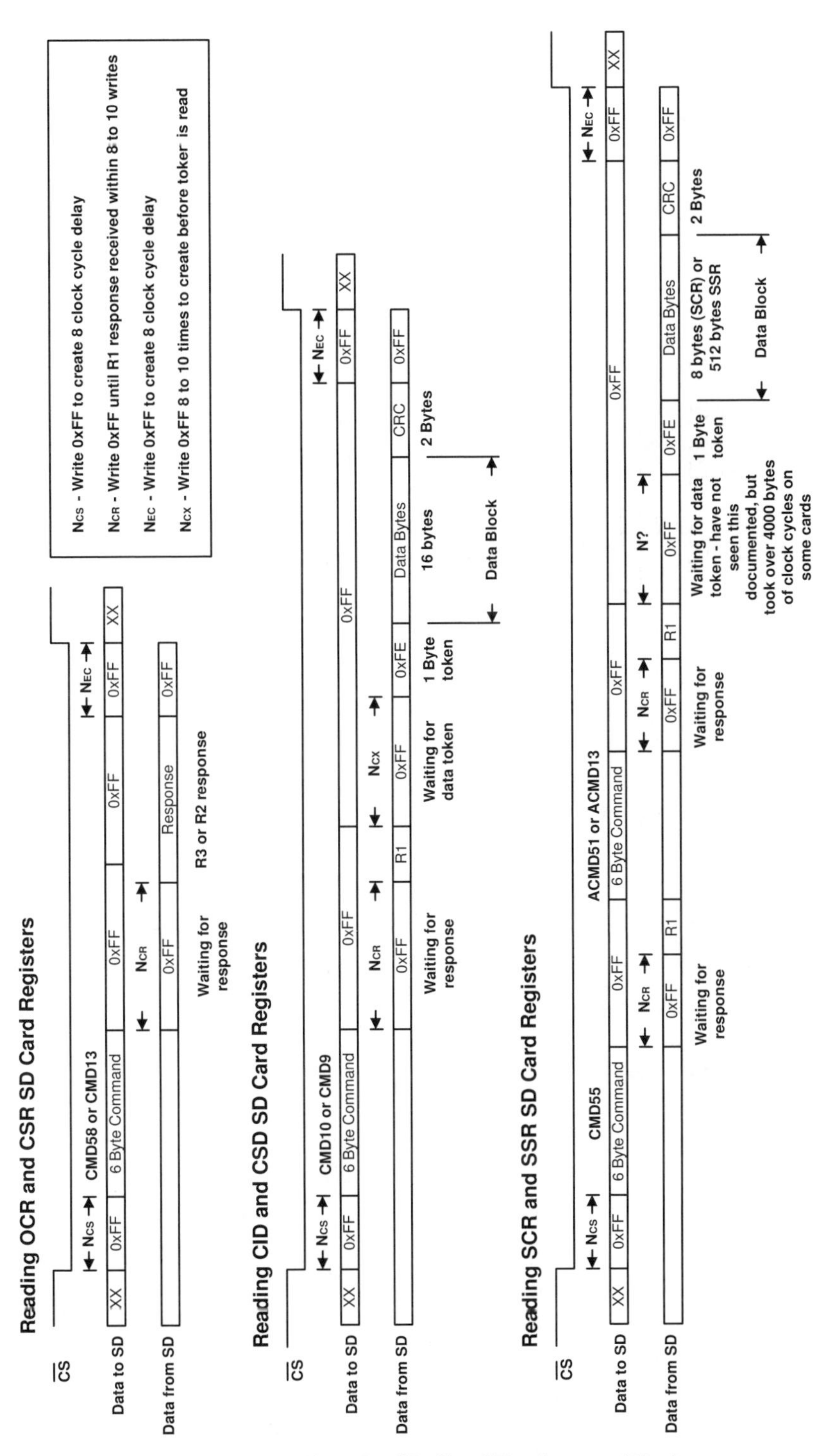

Figure 9.6: Reading the SD Card Registers – Timing

9.3.3 Reading and Writing the SD Card

Figure 9.7 shows the timing for single block read and write operations.

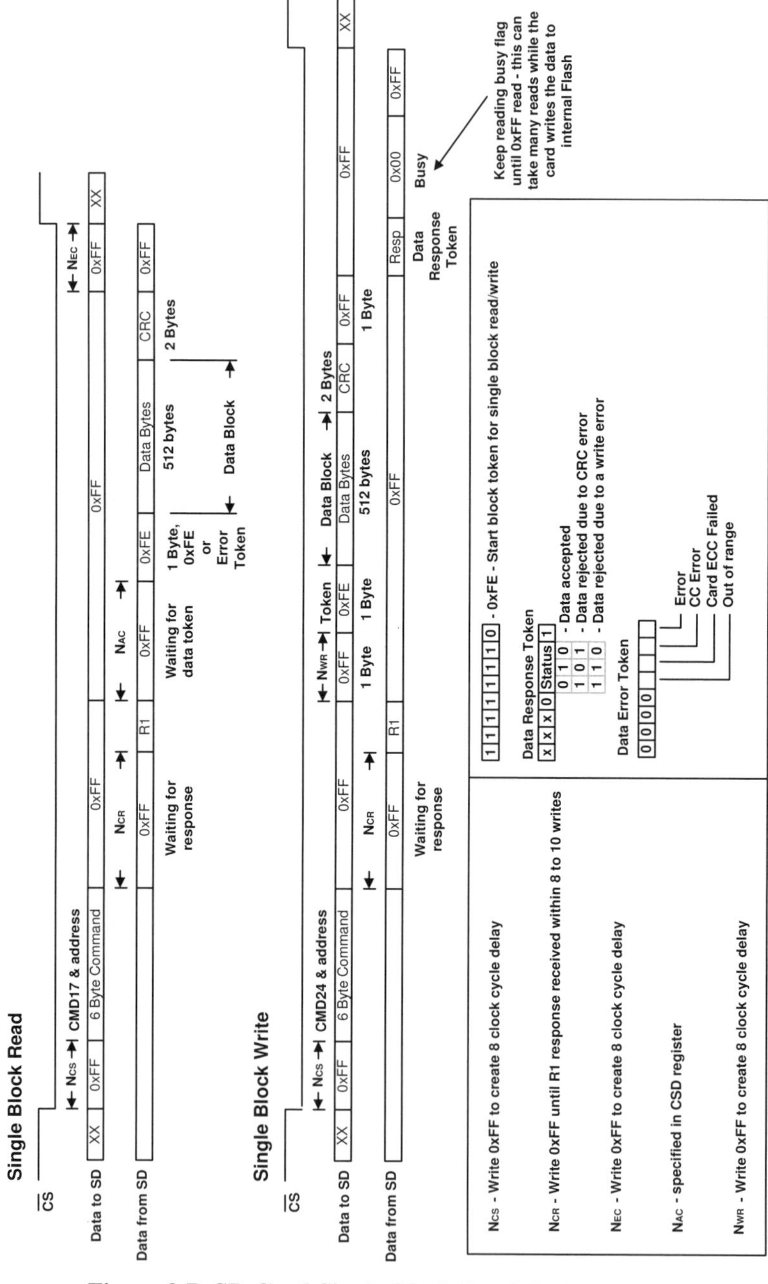

Figure 9.7: SD Card Single Block Read Write Timing

Reading a block of data

To perform a single block read, the read command is sent to the SD card (CMD17) the SD card will respond with an R1 response. The microcontroller must now send 0xFF multiple times until it receives back a start block token (0xFE) or an error token as shown in **figure 9.7**. The next byte after the token will be the first data byte of the block. To clock the data bytes into the microcontroller's SPI port, continue to send 0xFF bytes. After the data block has been clocked out of the SD card, the next two bytes will be a CRC checksum. One additional 0xFF byte must be sent to the SD card before pulling the /CS line high.

Writing a block of data

To perform a single block write, the write command (CMD24) must be sent to the SD card. An R1 response will be received back from the SD card. A single byte 0xFF must now be transmitted followed by the start block token (0xFE). The data bytes from the data block to be sent must directly follow the start block token. An additional two CRC bytes must be sent – 0xFF 0xFF can be sent if CRC checking has not been enabled. The second byte read back from the SD card after sending the last CRC byte should be response token from the SD card. The response token contains error/failure information to do with the write that has just been done as shown in **figure9.7**. The microcontroller will now read back bytes of value 0x00 while the card is busy writing the data block from its buffer to its Flash memory. It is possible for the microcontroller to de-assert the /CS signal and perform some other task and then come back and read bytes from the SD card again to see if it has finished writing. When finished writing, 0xFF will be read back instead of 0x00.

Data block sizes

High capacity SD memory cards have data block sizes fixed at 512 bytes. Standard capacity cards will have their data block sizes set to 512 bytes in the software written for this chapter in order to obtain maximum compatibility. Block length referred to in this chapter will always be 512 bytes for read and write operations.

9.3.4 SD Card Initialisation

After power has been applied to the SD card, it must be initialised by writing various commands to it. The commands must change it from its default SD card mode to SPI mode, check that it is voltage compatible with the host and send an initialisation command so that the card can initialise itself. During initialisation, it is also possible to check if an MMC card was inserted into the SD card socket. **Figure 9.8** shows a flowchart of what software must do in order to initialise the SD card in SPI mode.

SPI port configuration

Before the microcontroller can communicate with the SD card, it must initialise its SPI port. It is best to initialise this port at a low speed (maximum 400kHz) as this allows the initialisation software to detect if an MMC card was inserted in the SD card socket. Later, the speed can be increased up to 25MHz. If MMC card detection is not desired, the SPI port can be initialised at full speed.

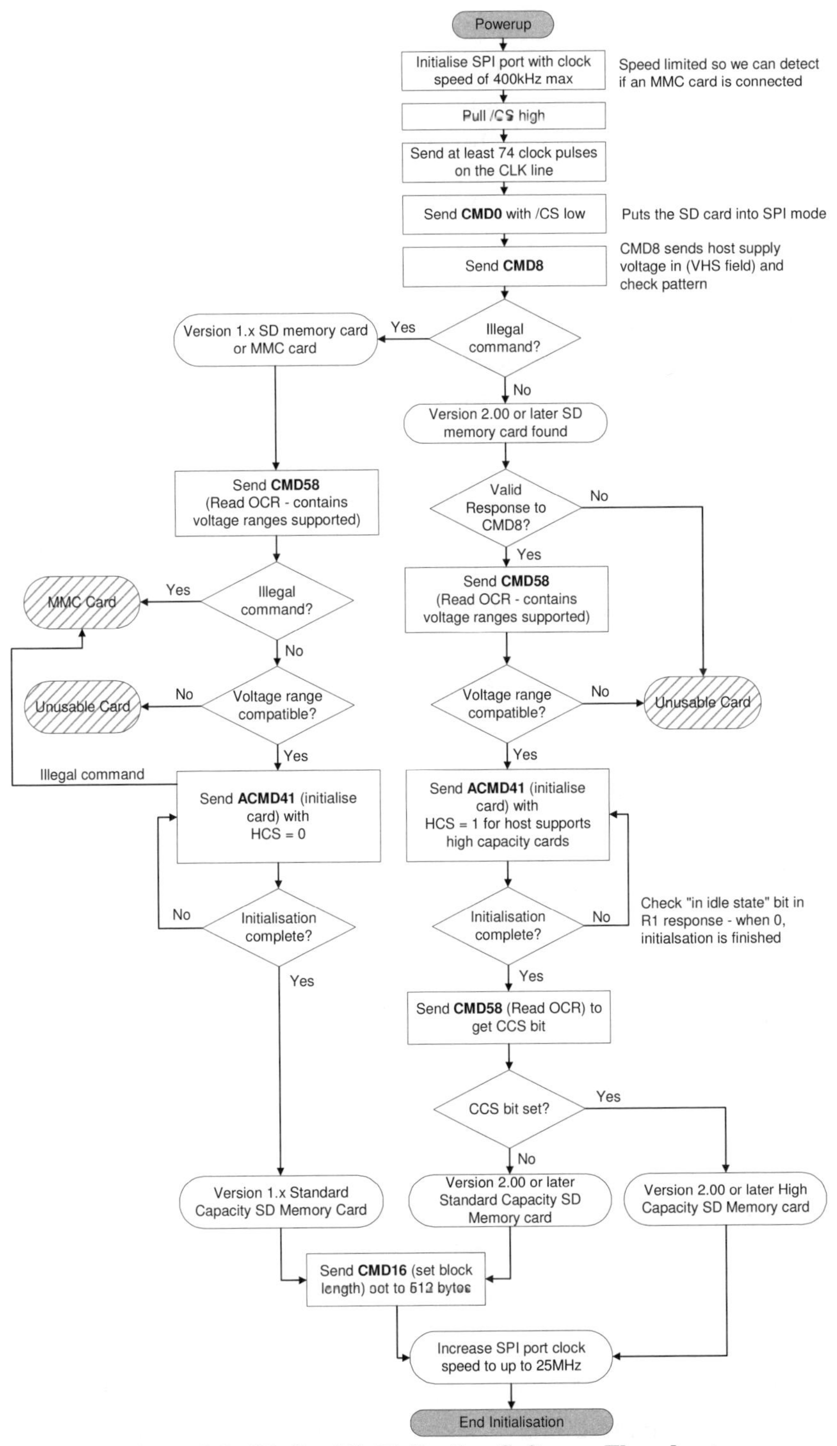

Figure 9.8: SD Card Initialisation Software Flowchart

After power-up and SPI port initialisation

After power has been applied to the SD card, the /CS line must be pulled high and at least **74 clock cycles** must be applied on the CLK (SPCK) line by the microcontroller. The SD card will now be in idle mode and will be waiting to communicate using the SD protocol.

Entering SPI mode

To change the card to SPI mode, **the /CS line must be asserted (pulled low) and a reset command (CMD0) sent** by the microcontroller. The card will respond with an R1 response. Once changed to SPI mode, the card will be stuck in SPI mode until the power to the card is switched off.

CRC checking

When sending the initial CMD0 command, the card is still in SD mode and a valid CRC checksum is required to be present in the command as CRC checking is enabled by default. After SPI mode has been entered, CRC checking is not enabled and commands can be sent with the CRC field blank except for CMD8 which must always have a valid CRC. In SPI mode, CRC checking can be enabled if desired by writing to one of the internal registers.

Initialisation process

The flowchart of **figure 9.8** shows the steps that must be taken to initialise the SD card. This has been implemented in the software programs in this chapter in the function `SDInit()` found in `sd.c`.

9.3.5 Miscellaneous Notes on SD Cards

MMC / SD cards

MMC cards and SD cards have many similarities and even have some of the same registers. Both can be initialised into SPI mode and use the same command response protocol. Many of the commands and responses are the same in the two card types. CMD1 can be used to initialise both SD can MMC cards instead of ACMD41. MMC cards do not recognise ACMD41.

An MMC card is thinner than an SD card and will fit into an SD card socket. The MMC card must first be communicated with at a clock speed of 400kHz or less. Once initialised, the MMC card can operate at up to 20MHz. A number of projects on the internet provide code that supports both SD can MMC cards on the same SPI port. The code that I have seen does not support SDHC cards.

If you want to support MMC code as well, this is not too difficult. I did not include MMC card code in this book as SD cards seem to be a lot more popular.

SD card data blocks
On SD cards, it is best to read and write data in 512 byte blocks and avoid reading or writing misaligned blocks (blocks that don't fall on a 512 byte boundary). Standard capacity cards can have a block length of as small as one byte. The default size can be read in one of the cards registers. High capacity cards have a fixed block length of 512 bytes.

Power control of the SD card
It may be desirable to be able to switch power to the SD card on and off from the microcontroller. This can be implemented by using a FET transistor controlled by a microcontroller output pin. Having control of the power is useful if the card goes into an unknown state as the microcontroller can then cycle the power. Power can also be saved by switching power to the SD card off when not used. When switching power to the SD card off, the pull-up resistors on the SPI port lines should be disabled to prevent feeding power into the card on these lines.

9.4 SD Card Software – Initialisation, Reading and Writing

The programs in this section have been tested with version 2.00 standard capacity, version 2.00 high capacity and version 1.x standard capacity SD memory cards.

Three software programs are presented here and build up from a simpler program that initialises the SD card to reading and writing the SD card. Starting with a simpler program will help you to understand the software better without the clutter of extra functions needed only in later programs.

The first program initialises the SPI port and SD card and uses a single function to read two of the SD card registers. The second program reads all of the SD card registers and interprets some of the contents of the registers in order to display information such as the date of manufacture of the SD card. The third program performs reading and writing of SD card sectors.

9.4.1 Hardware Resources Used

All the programs in this section and the FAT program in section **9.5** use the following hardware resources:

- **DBGU port** for sending information to be displayed on a terminal emulator.
- **SPI port** pins MOSI, MISO and SPCK.
- **GPIO pin PA11** used as /CS line for SPI port and manually toggled in software.

SPI hardware is set up as per the left circuit in **figure 9.2** and it is assumed by software that the SD card is connected to the SPI port at power-up.

9.4.2 Initialisation of the SD Card

Program name
`SD_card_init_RAM`

What it does
This program initialises the SPI port and then the SD card. Results of the initialisation are sent out of the DBGU serial port for display on a terminal emulator (success or failure). The contents of the SD card CSD and CID registers are also read and sent for display.

Program structure
`spi.c`
Contains functions for initialising the SPI port and sending and receiving bytes.

`SPIInit()` – Enables SPI port pins MOSI, MISO and SPCK to be assigned to their peripheral function. Enables PA11 as a GPIO output pin for use as the /CS signal. Starts the SPI port clock at low speed.

`SPIWriteByte()` – Sends a single byte on the SPI port and polls the SPI port to wait for the byte to be transferred to the SPI port shift register.

`SPIReadByte()` – Writes a byte (0xFF) in order to clock a byte into the SPI port, polls a flag to see that the byte has been received, then reads a single byte from the SPI port.

sd.c

SDInit() – Calls **SPIInit()** to initialise the SPI port and then initialises the SD card. Returns 0 if initialisation succeeded otherwise an error code. Before returning, it increments the SPI port clock speed to a higher speed.

SDCommand() – Sends the 6 byte SD command with the command number and argument fields sent to the function as arguments.

There are two functions for receiving responses from the SD card – **SDGetR1Response()** gets the 8-bit R1 response and **SDGetR3R7Response()** gets the 32-bit part of either the R3 or R7 response. **SDGetR1Response()** must be called before **SDGetR3R7Response()** as these response packets are always preceded with and R1 response.
The **SDGetCID_CSDRegister()** function gets the contents of either the CID or CSD 128 bit (16-byte) registers.

dbgu.c

Two new functions have been added to **dbgu.c** in order to send decimal and hexadecimal numbers for display on a terminal emulator without the need for using **sprintf()** which uses a lot of memory.
DBGUTxNum() converts a number to decimal ASCII and then sends it out of the DBGU port.
DBGUTxHexNum() converts a single byte to a hexadecimal ASCII number and sends it out of the DBGU port.

How the program works

Before running the program, the SD card must be plugged into the SD card socket that is interfaced to the microcontroller. This is necessary as the program initialises the card at the start of the program so does not support hot plugging of the card.

SDInit() is called at the beginning of the program to initialise the SPI port and the SD card. This function initialises the SD card according to the information in the SD card simplified specification version 2.00. If an error occurred when trying to initialise the SD card, the program will send the error code to the terminal emulator and then do nothing more. This function also provides information about the SD card that was discovered during initialisation and saves it in a variable whose address is passed to it. The information includes whether the SD card is a version 1.x or 2.00 card and whether the card is standard or high capacity.

If card initialisation was successful, the program reads the CSD and CID registers and displays these numbers on the terminal emulator in hexadecimal.

Making changes to the program
To change the **GPIO pin used for the /CS line** on the SD card from PA11, change the value of `PIO_SD_CS_PIN` defined in `spi.h`. e.g. to change to PA0, change the value to `0x00000001`. The bit set in the 32 bit word from bit position 0 to 31 corresponds to port pins PA0 to PA31.

The **SPI port clock speed**, can be changed at the end of the `SDInit()` function. When the card is initialised, it is run at a low clock speed initially set in `SPIInit()`. If your card initialises correctly, but experiences problems reading registers and in later programs, reading or writing data, try lowering the clock speed in `SDInit()`.
The timeout value waiting for the card to initialise is set to 32000 in a `for` loop in `SDInit()`. This could also be a possible problem area if the card does not initialise properly, although this number is very high and I wouldn't expect a problem to be encountered here.

9.4.3 Reading and Interpreting SD Card Registers

Program name
`SD_card_registers_RAM`

What it does
This program displays the contents of the following SD card registers on a terminal emulator: CSD, CID, OCR, SCR, SSR and CSR. It then interprets some of the information in some of the registers and displays the results. The meanings of the bits in these registers can be found in the SD card simplified specification version 2.00.

Program structure
This program was derived from the previous program and then modified.
The following functions were added to `sd.c` to get the contents of various SD card registers:
`SDGetOCRRegister()` – Gets the contents of the OCR register
`SDGetSCRRegister()` – Gets the contents of the SCR register
`SDGetSSRRegister()` – Gets the contents of the SSR register
`SDGetCSRRegister()` – Gets the contents of the CSR register

The following functions were added to main.c to interpret some of the bits in some of the SD card registers:
`DisplayCSDRegisterInfo()` – Displays the CSD register version information, and the card size in bytes. Note that the card size in bytes will only work for cards with a maximum size of 4GB as the size is stored in a 32 unsigned integer which can store a maximum value of 4GB (2^{32}).
`DisplayCIDRegisterInfo()` – Displays the date that the card was manufactured.
`DisplaySCRRegisterInfo()` – Displays the version of the SD card physical layer specification.
`DisplaySSRRegisterInfo()` – Displays the speed class of the SD card.

These functions are only displaying a small portion of the information available in the SD card registers. More on this information is available in the SD card specification document.

How the program works
The program works exactly the same as the previous one, but calls the extra functions listed above to display the contents and information in the additional registers.

9.4.4 Raw Writing and Reading Data

Warning: Raw writing may remove formatting information from the card and corrupt any data that is on the card. Don't use a SD card with this program if it contains valuable data that has not been backed up.

The write function has been commented out of the code in the example code that you have downloaded in order to prevent accidental corruption of data on the SD card. The program will only perform the read function. You will need to uncomment the write function and then build the program in order for it to write to the SD card.

Program name
`SD_card_raw_rw_RAM`

What it does

The program first displays the contents of the CSD and CID registers on a terminal emulator. It then reads a 512 byte block of data from the SD card (sector 100). A 512 byte array is then filled with a count value and 0xAAAA5555 fills the first four bytes. This data block is then written to sector 100 of the SD card. In order to see the written data displayed, the program must be re-run.

Program structure

This program was derived from the `SD_card_init_RAM` program (the first program in this section. The following modifications were made to the program:

<u>sd.c</u>

Added function **`SDSetSectorSize512()`** which sets the sector size of standard capacity memory cards to 512 bytes. This function is then called at the end of **`SDInit()`** if a standard capacity memory card is detected.
Added functions:
`SDReadSector()` – reads a 512 byte sector from the SD card at the specified sector number passed to it as an argument.
`SDWriteSector()` – writes 512 bytes of data to a sector on the SD card specified in the argument passed to it.

<u>main.c</u>

Added two global arrays of 512 bytes each to use as read and write buffers. Added calls to SD card read and write functions.

How the program works

The program calls **`SDReadSector()`** to read sector 100 of the SD card and store its contents in the global buffer **`buf_read512[]`**. It then displays this information on a terminal emulator.

The global buffer **`buf_write512[]`** is then filled with a count value. **`SDWriteSector()`** is called to write the global value to sector 100.

The next time that the program is run, the data previously written will be displayed on the terminal emulator.

9.5 FAT File System

The **FAT (File Allocation Table)** file system can be read by PCs and implemented on embedded systems. Most SD cards will come pre-formatted with the FAT file system. This section describes only the basics of the FAT file system. I have chosen rather to port FAT file system code that has been made available for free on the internet. This will be described in the next section. A document is available which describes the details of how the FAT file system works. It is available here:
www.microsoft.com/whdc/system/platform/firmware/fatgen.mspx

You will need to accept the license agreement in order to download the document.

There are three types of FAT file systems, each one being added to allow addressing of more data as data storage devices got bigger:

FAT12 – size limit of up to 32MB
FAT16 – size limit of up to 2GB
FAT32 – size limit of up to 2TB

FAT file systems divide the storage device into four regions:
- **Reserved region** – contains boot sector
- **FAT region** – contains the File Allocation Table itself, usually two copies for redundancy/reliability
- **Root directory region** (not present on FAT 32) – information on directories and files stored in the root directory (such as filenames, dates, times, etc.). FAT32 stores this information in the data region.
- **File and directory data region** – files and directories are stored here.

Data is stored as sectors and clusters in a FAT file system:

Sectors
The FAT file system groups bytes on the storage device into sectors. Sectors can be 512, 1024, 2048 or 4096 bytes long. 512 byte long sectors are most common.

Clusters
Sectors are grouped into clusters. Clusters can consist of 1, 2, 4, 8, 16, 32, 64 or 128 sectors.

A sector/cluster combination (sector size × cluster size) must never be greater than **32k bytes**. So (4096 × 8), (2048 × 16), (1024 × 32) are the maximum size sector/cluster size combinations allowed. The specification document reference above says that a cluster size of 128 is valid, but combining 128 with the minimum sector size of 512 gives a combination greater than 32k. Other references say that **64k** is a valid combination.

FAT

The File Allocation Table after which the FAT file system is named is a map of the data region of the file system. Each entry in this table represents one cluster in the data region. The sizes of the entries in the FAT are 12 bits for FAT12, 16 bits for FAT16 and 32 bits for FAT32. Each entry in the FAT tells us if a cluster is available for data storage or already has data in it. If a file stored on the device is larger than the size of a cluster, it will take up two or more clusters. These clusters may not necessarily be consecutive clusters. In this case the FAT acts as a linked list and tells us which cluster number is the next one that holds the next part of the file.

FAT size limits

FAT12 is limited to a maximum number of clusters of 4084. The disk's size is stored as a 16 bit count of 512 byte sectors which limits its size to 32MB (2^{16} × 512).
FAT16 can have a maximum of 65524 clusters of 32k bytes each. Disk maximum size is therefore 65524 × 32k = about 2GB.
FAT32 has 32 bit FAT table entries, but only 28 bits are used for addressing clusters. This would allow a maximum disk size = 2^{28} × 32k = 8TB, but the boot sector uses a 32 bit number for sector count which limits the size to 2TB (2^{32} × 512 = 2TB).

Number of clusters for data storage for FAT12 and FAT16 are not 4096 (2^{12}) and 65536 (2^{16}) as some of the space is reserved for use by the file system.

9.6 SD Card Software – FAT File System Implementation

9.6.1 About the FAT File System Software

In this section we will port the **FatFs** FAT file system code from **elm-chan.org** to our embedded system so that we can read and write to an SD card that is formatted with the FAT file system.

FatFs is a generic FAT file system module for embedded systems from the ELM website (elm-chan.org). It is available free of charge as C source code. FatFs can be downloaded at this page: **elm-chan.org/fsw/ff/00index_e.html**
Scroll down to the bottom of the page and look under the "Resources" heading. At the time of writing the newest version of FatFs was FatFs R0.07e and was downloaded as a zipped file called **`ff007e.zip`**.

FatFs main features

- ANSI C compliant
- Hardware independent
- Small code footprint
- Supports FAT12, FAT16 and FAT32 formatted storage devices

FatFs consists of the following source code files found in the **`src`** directory in **`ff007e.zip`**:

`diskio.c` and **`diskio.h`** – low level functions to access the storage device such as an SD card. This file must be modified to supply the low level functions that the system that it is running on needs to access the storage device for initialising, reading and writing.

`ff.c` and **`ff.h`** – the FAT file system code.

`ffconf.h` – configuration file for changing options before compiling the code. For example FatFs can be compiled to access storage media as read only or to exclude some functions to make it smaller.

`integer.h` – definitions used in the FatFs code.

9.6.2 Porting the FAT File System Code

FatFs is easy to port to our embedded system. Code must be added to the **diskio.c** file of FatFs, but this is code that we have previously written to read and write the SD card.

Details of the example program for this section are shown below, followed by an explanation of how the FatFs code was ported to create this program. This program was derived from our previous program (SD_card_raw_rw_RAM) so uses all of the same hardware resources, but adds FAT file system functionality to it and does not do raw reading or writing.

Program name
SD_card_FAT_RAM

What it does
The program initialises the SD card and then mounts the file system. The contents of the root directory of the FAT formatted SD card are then displayed on a terminal emulator. A new empty file called **new.txt** is created in the root directory. A new empty file called **text.txt** is created in the **Test** directory. This directory must exist on the SD card for the file creation to work. If it does not, the program will continue to run without generating an error or creating the file. A file called **story.txt** is opened and the first 512 byte sector read and displayed on a terminal emulator. This file must exist and have some text in it in order for this function to work. Next a file called **info.txt** is created in the root directory and some text written to it. Finally a directory called **temp** is created in the root directory.

The file and directory needed for some of the functions in this program to work need to be added by connecting the SD card to a PC and creating them. If your PC does not have an SD card slot, you can buy a multi-card reader USB device for this purpose.

The idea of this program is to test some of the functions from the FatFs code so that you can use them in your own programs. The functions that write to the disk have been commented out in the source code as a precaution, so you need to uncomment them before the program will work as described.

Program structure

The program is a copy of the previous program, **SD_card_raw_rw_RAM**, with the files from the FatFs **src** directory added. The porting process is described below.

Steps taken to port the FatFs code

1) **Create Project:** A copy of our previous program, **SD_card_raw_rw_RAM**, was made and renamed to **SD_card_FAT_RAM**.
2) **Add FatFs Code:** The contents of the **src** file in **ff007e.zip** were copied to the project's **Code** directory.
3) **Modify make file:** The make file of the project was modified to change its name and to compile and link the **ff.c** and **diskio.c** files from the FatFs code
4) **Increase Stack Size:** The stack size was increased in **startup.s** from 1k bytes to 4k bytes as a precaution, although it looks like the stack size only needs to be increased if long file name support is to be used, which is not the default setting.
5) **diskio.c modified: sd.h** included at the top of the file.
6) **diskio.c modified:** All functions in this file were made to return 0 except **disk_read()** and **disk_write()**.
7) **diskio.c modified: disk_read()** calls **SDReadSector()** and **disk_write()** calls **SDWriteSector()** to do low level reading and writing.
8) **diskio.c and diskio.h modified:** Added function **get_fattime()** needed by FatFs for time and date stamping files and directories when they are written. As my system does not have a real-time clock, this function simply returns 0.
9) **main.c modified:** added a call to **f_mount()** to mount the file system, followed by the code to perform the functions described previously – read root directory, create new file, etc.

10. Advanced ARM Microcontrollers

The AT91SAM7S256 is a great stand-alone microcontroller to start with and can be used in many applications; however a larger project may require more I/O pins, an external bus or even a faster processor. This chapter provides information on more advanced ARM microcontrollers from Atmel that you may be interested in. Information is also provided on the evaluation boards for these microcontrollers that I have personally tested.

Although there are many microcontroller manufacturers that use ARM cores in their microcontrollers and you are free to choose any manufacturer, some of the advantages of sticking with Atmel microcontrollers are:

- All of these microcontrollers can be debugged using the same Atmel AT91SAM-ICE JTAG emulator that you may have purchased for use with the AT91SAM7S-EK, so there is no need to buy additional expensive tools.
- C programs can be written for all these microcontrollers using free GNU ARM software tools and SAM-BA used to load the programs.
- There are similarities between Atmel microcontrollers across the range that will help to decrease the time to learn a new microcontroller, e.g. all the microcontrollers discussed here have the DBGU serial port.

There are many more Atmel ARM microcontroller families available than are discussed here, but the microcontrollers discussed have been selected from across the range to give you a good overview of what is available:

- The **AT91SAM7SE** – An **ARM7** microcontroller with external bus interface. Maximum speed 55MHz.
- The **ATSAM3U** – An **ARM Cortex-M3** microcontroller with external bus. One of the latest offerings from Atmel. Maximum speed 96MHz.
- The **AT91SAM9263** – An **ARM9** microcontroller with numerous on-chip peripherals – LCD controller, camera interface, Ethernet, USB host ports and more. Maximum speed 240MHz.

10.1 AT91SAM7SE – ARM7

10.1.1. AT91SAM7SE Features

The AT91SAM7SE range is very similar to the AT91SAM7S microcontroller range, the main difference being that the AT91SAM7SE has an external bus that can interface to external memory and peripherals. If the external bus is not used, the pins that make up the external bus can be used as general purpose I/O pins or connected to the internal peripherals that the pins are multiplexed with.

The AT91SAM7SE has the following main features:

- ARM7TDMI Core

- Three microcontrollers in the range, differing only in on-chip memory sizes:
  - AT91SAM7SE512 512k bytes Flash 32k bytes SRAM
  - AT91SAM7SE256 256k bytes Flash 32k bytes SRAM
  - AT91SAM7SE32 32k bytes Flash 8k bytes SRAM

- External Bus Interface (EBI) capable of interfacing to:
  - SDRAM
  - SRAM
  - Compact Flash
  - NAND Flash

- Three 32 bit ports that are multiplexed with on-chip peripheral devices:
  - Port A: PA0 to PA31
  - Port B: PB0 to PB31
  - Port C: PC0 to PC23 (only 24 pins of port C are available on external pins)

To get a good idea of what the AT91SAM7SE features are, a comparison is made with a microcontroller that you should be familiar with by now, the AT91SAM7S. **Table 10.1** compares the AT91SAM7SE256 with the AT91SAM7S256. Remember that the other AT91SAM7SE family members only differ from each other in the size of on-chip memory.

The external data bus is capable of 16-bit or 32-bit wide transfers. The address bus can be up to 23 bits wide with up to 8 chip select lines.

At 3.3V, the I/O pins can source or sink the following:
PA0 – PA3, 16mA each
PA4 – PA31, 8mA each
PB0 – PB31, 8mA each
PC0 – PC23, 8mA each
NRST pin, 8mA

A maximum total current of 200mA can be drawn from all I/O pins simultaneously.

AT91SAM7SE microcontrollers are good way to upgrade from the AT91SAM7S microcontrollers if you require any of the following features:

- Your project requires more I/O pins than the AT91SAM7S has.
- You want to interface a hardware device to a bus on the microcontroller.
- The software that you are running requires extra memory for programs and/or data storage in the form of SDRAM, SRAM and/or Flash on an external bus.

The advantages of moving to the AT91SAM7SE from the AT91SAM7S are:

- The ARM cores and many of microcontroller features are identical, so learning the new microcontroller requires minimal effort.
- You can use the same toolchain for both microcontrollers.
- You can use the same JTAG ICE for both microcontrollers.

FEATURES	AT91SAM7SE256	AT91SAM7S256
A. Differences		
SRAM	32k bytes	64k bytes
External Bus	1	-
I/O Pins	88	32
MPU	Memory Protection Unit	-
Packages	LQFP128	LQFP64
	LFBGA144	QFN64
B. Similarities		
ARM Core	7TDMI (ARM7 Core)	
Flash	256k	
RTT	1	
10-bit ADC Channels	8	
ADC pins	AD4 to AD7 are dedicated ADC pins – not multiplexed	
DMA Channels	11	
USART	2	
DBGU	1	
SPI	1	
TWI	1	
SSC	1	
USB Device Port	1 Full Speed	
PWM Controllers	4	
High Current Pads	4	
16-bit Timers	3	
PIT	1	
WDT	1	
Power On Reset	1	
Brown Out Detection	1	
RC Oscillator	1	
Crystal Oscillator/PLL	1	
I/O Voltage	3.3V	
In-System Programming	Yes	
Single Supply	Yes	

Table 10.1: Comparison of the AT91SAM7SE256 and AT91SAM7S256 Microcontrollers from Atmel

10.1.2 AT91SAM7SE Evaluation Kit

The AT91SAM7SE family of ARM7 microcontrollers from Atmel can be evaluated using the AT91SAM7SE-EK evaluation kit. **Figure 10.1** shows the AT91SAM7SE-EK board.

The kit has the following contents:

- AT91SAM7SE-EK circuit board
- 5V power supply
- A/B type USB cable
- RS-232 cable with two 9-pin D-type female connectors (crosses tx and rx lines for connection to a PC.
- CD-ROM

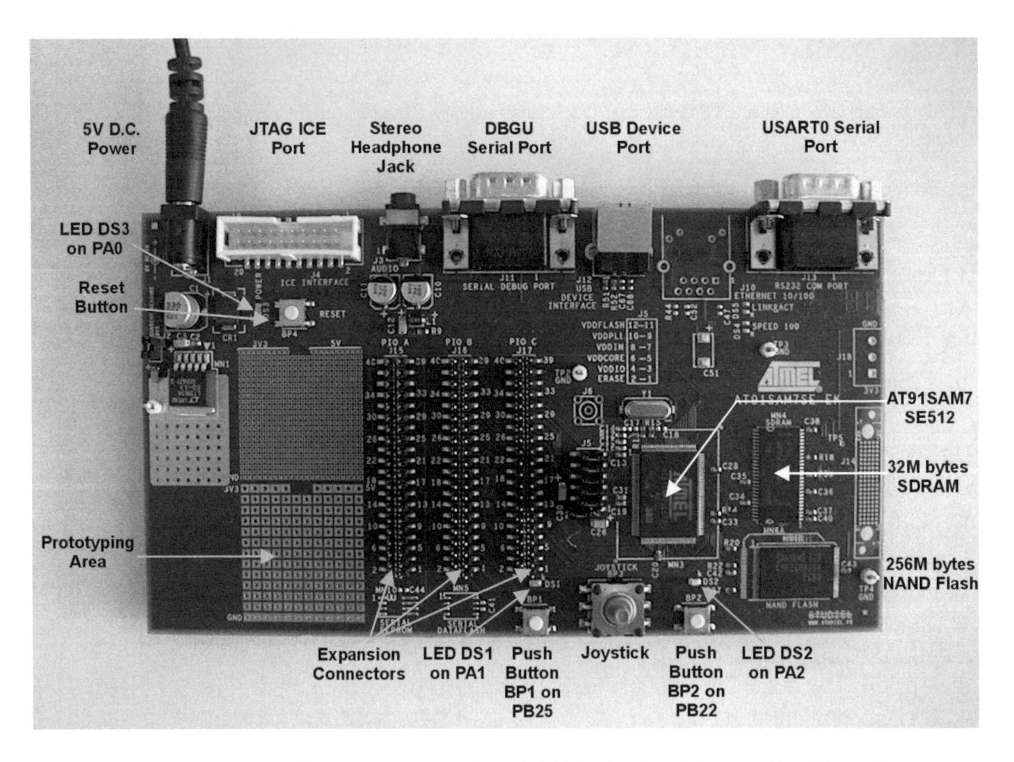

Figure 10.1: The Atmel AT91SAM7SE-EK Evaluation Board

The AT91SAM7SE-EK board has the following features:

- AT91SAM7SE512 ARM7 microcontroller in a 128 pin LQFP package
- 32M bytes SDRAM memory with 16-bit data bus
- 256M bytes of NAND Flash memory with 8-bit data bus
- A USB device port
- Two RS-232 serial ports (1 DBGU port, one additional serial port (USART0))
- JTAG ICE debug interface (20 pin IDC connector)
- Stereo audio DAC connected to a 3.5mm headphone jack
- One power LED, two general purpose LEDs
- A Joystick
- Two push buttons
- A reset push button
- Three expansion connectors for ports PIOA, PIOB and PIOC (40 pin headers – 2 rows × 20 pins each)
- An 18.432MHz crystal for the microcontroller clock

The AT91SAM7SE-EK board must be powered with external 5V and does not allow for powering from the USB port. A 5V power supply is supplied with the kit. The centre pin of the power connector is positive.

10.1.3 AT91SAM7SE Programming

Example programs for the AT91SAM7SE-EK can be downloaded from the Atmel website (**www.atmel.com**). Download the **`at91sam7se-ek.zip`** file.

Case study – compiling and running the `basic-pwm-project-at91sam7se-ek` example program:

1) Obtaining the project:
When unzipped, the **`at91sam7se-ek.zip`** file produces a directory called **`at91sam7se-ek`**. The project that we are interested in is found at the following location in this directory:
`at91sam7se-ek\packages\basic-pwm project-at91sam7se-ek-gnu.zip`
Unzip this file and the following directory is produced that contains the project:
`basic-pwm-project-at91sam7se-ek`

2) Compiling the project:

To compile the project, you will need the YAGARTO toolchain or the CodeSourcery G++ toolchain loaded on your computer. The binary outputs of the compiled project are already available in the unzipped project, but we will clean the project and compile from scratch to make sure that our toolchain is working so that we can build our own projects.

In Windows, navigate to: **basic-pwm-project-at91sam7se-ek\basic-pwm-project**

Open the make file for editing.

The following lines at the top of the make file had to be changed to reflect the chip and board in use, i.e. the AT91SAMSE, as the default make file was set up for a different chip and board:

```
CHIP  = at91sam7se256
BOARD = at91sam7se-ek
```

The make file is also set up to use the CodeSourcery G++ toolchain by default. If you are using YAGARTO, change the following line:

```
CROSS_COMPILE = arm-none-eabi-
```

to:

```
CROSS_COMPILE = arm-elf-
```

Now start a command prompt in Windows:

start->Run..., type **cmd** in the **Open:** box and hit **Enter**.

Change to the **basic-pwm-project-at91sam7se-ek\basic-pwm-project** directory in the command prompt window. (Easiest way is to type **cd**<space> in the window, copy the path out of Windows explorer, right click on the **cmd** box, select **Edit->Paste** from the pop-up menu, hit **Enter**).

In the **cmd** window, type **make clean** and hit **Enter**. Now type `make` to build the project from scratch. If the build was successful, the **bin** sub-directory of the **basic-pwm-project** directory will contain the binary files compiled for Flash, SRAM and SDRAM memory.

3) Loading the program to the board:

When all Flash is erased by closing jumper J5-1 on the AT91SAM7SE-EK board (i.e. pulling the ERASE pin high) and powering up the board, then powering down the board and opening jumper J5-1, the next time that the board is powered up, it will run the internal SAM-BA boot program. To be able to boot from Flash memory again, the Boot from Flash (GPNVM2) bit must be set using the SAM-BA PC application. Take note of this difference if you are used to the AT91SAM7S which overwrites the boot loader in Flash memory when a user program is loaded.

To load and run the program from SDRAM using either the USB port or the Atmel SAM-ICE, do the following:

- Connect to the board using the SAM-BA PC program using either the USB connection or JTAG connection with SAM-ICE.
- Click the SDRAM tab.
- Click the **Execute** button to run the "**Enable SDRAM**" script.
- Open the **basic-pwm-project-at91sam7se-ek-at91sam7se256-sdram.bin** file using the file open icon next to the **Send File** button.
- Click the **Send File** button.
- Type "go 0x20000000" at the prompt in the box at the very bottom of the SAM-BA program and hit **Enter**.

The program should start running. It uses the PWM channels connected to LEDs DS1 and DS2 to alternately dim each LED.

10.2 ATSAM3U – ARM Cortex-M3

10.2.1 ATSAM3U Features

The ATSAM3U has the following main features:
- ARM Cortex-M3 Core
- There are six microcontrollers in the ATSAM3U range (ATSAM3U**4E**, ATSAM3U**2E**, ATSAM3U**1E**, ATSAM3U**4C**, ATSAM3U**2C** and ATSAM3U**1C**). The range is divided into two halves: The E type (ATSAM3UxE) and the C type (ATSAM3UxC). The E type are packaged in 144 pin packages and have 96 I/O pins, the C type are packaged in 100 pin packages and have 57 I/O pins. Memory is arranged as follows:
  - ATSAM3U4x 256k bytes Flash 52k bytes SRAM
  - ATSAM3U2x 128k bytes Flash 36k bytes SRAM
  - ATSAM3U1x 64k bytes Flash 20k bytes SRAM
  - (where x is the E or C type)

- External Bus Interface (EBI) (8 or 16 bit on E type, 8 bit on C type), capable of interfacing to:
  - SRAM
  - NAND Flash

- Three 32 bit ports that are multiplexed with on-chip peripheral devices:
  - Port A: PA0 to PA31
  - Port B: PB0 to PB31, (PB0 to PB24 on C type)
  - Port C: PC0 to PC31 (not present on C type)

Figure 10.2: The ATSAM3U-EK Evaluation Board

FEATURES	ATSAM3U4E	AT91SAM7S256
A. Differences		
ARM Core	Cortex-M3	7TDMI
SRAM	52k bytes	64k bytes
External Bus	1	-
RTC (Real Time Clock)	1	-
12-bit ADC Channels	8	-
DMA channels	17	11
Max. Clock Speed	96MHz	55MHz
I/O Pins	96	32
USART	4	2
SPI	5	1
TWI	2	1
MCI	1	-
USB Device Port	High Speed (HS)	Full Speed (FS)
High current pads	-	4
On-chip RC Oscillator	2	1
Crystal Oscillator/PLL	2	1
MPU	Memory Protection Unit	-
Packages	LQFP144	LQFP64
	BGA144	QFN64
B. Similarities		
Flash	256k	
10-bit ADC Channels	8	
DBGU	1	
SSC	1	
PWM Controllers	4	
16-bit Timers	3	
RTT	1	
PIT	1	
WDT	1	
Power On Reset	1	
Brown Out Detection	1	
I/O Voltage	3.3V	
In-System Programming	Yes	
Single Supply	Yes	

Table 10.2: Comparison of the ATSAM3U4E and AT91SAM7S256 Microcontrollers

The external data bus is capable of 8-bit or 16-bit wide transfers on the E type and 8-bit transfers only on the C type. The address bus is 24-bits wide with 4 chip selects on E type, 8-bit with 2 chip selects on C type.

At 3.3V, the I/O pins can source/sink the following currents:

I/O Pins	Source	Sink
PA3, PA15	15mA	9mA
All other pins	3mA	6mA

A maximum total current of 100mA can be drawn from all I/O pins simultaneously on the 100 pin packages and 130mA on the 144 pin packages.

Table 10.2 compares the ATSAM3U4E with the AT91SAM7S256.

10.2.2 ATSAM3U-EK Evaluation Kit

The ATSAM3U family of ARM Cortex-M3 microcontrollers from Atmel can be evaluated using the ATSAM3U-EK evaluation kit. **Figure 10.2** shows the ATSAM3U-EK board.

The kit has the following contents:

- ATSAM3U-EK circuit board
- 5V power supply
- CR1225, 3V lithium battery
- A/B type USB cable
- RS-232 cable with two 9-pin D-type female connectors (crosses tx and rx) lines for connection to a PC.

The ATSAM3U-EK board has the following features:

- ATSAM3U4E ARM Cortex-M3 microcontroller in a 144 pin LQFP package
- 1M byte PSRAM (pseudo-static RAM), 16-bit data bus
- 256M bytes of NAND Flash memory with 8-bit data bus
- Backup battery
- 2.8 inch TFT colour LCD display with touch-panel and backlight
- A USB device port (high speed)
- Two RS-232 serial ports (1 DBGU port, one additional serial port)
- JTAG ICE debug interface (20 pin IDC connector)
- Audio codec with input and output jacks (3.5mm): stereo headphone out, stereo line in, mono microphone in
- SD/MMC card socket
- 3-D accelerometer sensor
- Temperature sensor
- 2 user buttons marked left and right
- One power LED, two general purpose LEDs
- On-board power regulation with shutdown control (by the SAM3 chip)
- BNC connectors for ADC inputs
- Reset and wakeup buttons
- User potentiometer connected to the ADC input
- ZigBee® connector
- Three expansion connectors for ports PIOA, PIOB and PIOC (40 pin headers – 2 rows × 20 pins each)
- A 12MHz crystal for the microcontroller clock
- 32.768kHz crystal

The ATSAM3U-EK board must be powered with external 5V and does not allow for powering from the USB port. A 5V power supply is supplied with the kit. The centre pin of the power connector is positive.

10.2.3 ATSAM3U Programming

The example programs provided with the ATSAM3U-EK can be compiled with the free open-source "Sourcery G++ Lite for ARM EABI". Download and installation instructions are provide on the ATSAM3U-EK page at the Atmel website. Example programs can be downloaded from Atmel and SAM-BA used to program the microcontroller. The SAM-ICE JTAG emulator can be used to debug the board.

The default program that runs when the board is booted up for the first time displays various options (as icons) on the touch LCD display. Each option can be selected by touching an icon on the LCD display. This should give you some idea of the capabilities of this microcontroller.

The following can be selected from the menu:

- **USB flash icon** – the board can be plugged into a PC with a USB cable and the contents of the NAND Flash memory will appear as a drive on the PC. The drive contains a tutorial on how to start using the evaluation kit (in html format), a demo program and a WAV file. If a card is inserted in the SD card socket, it will also appear as a drive on the PC.

- **Music icon** – will play the WAV file that is in the NAND Flash.

- **Power plug icon** – demonstrates various power down modes.

- **I.C. icon** – displays a power-point type presentation that flips through various slides that contain information on the ATSAM3U microcontrollers.

- **Tool icon** – allows date, time, LCD brightness, thermometer and slide presentation duration to be set.

10.3 AT91SAM9263 – ARM9

10.3.1 AT91SAM9263 Features

The AT91SAM9263 has a most impressive array of on-chip peripherals and at the time of writing had the highest number of peripherals of any of the ARM microcontrollers from Atmel. It is suited to high end applications such as portable multimedia devices and handheld GPS.

This microcontroller has no on-chip Flash memory and is available only in a BGA package (with 324 balls!) making it less suitable to hobbyists, however at the time of writing, a new range of ARM9 microcontrollers from Atmel were available as samples – the AT91SAM9XE with the same ARM9 core as the AT91SAM9263. These devices have on chip Flash memory and will be available in PQFP 208 packages as well as BGA packages.

The AT91SAM9263 has the following main features:

- ARM926EJ-S Core with these main features:
  - DSP Instruction Extensions, Jazelle® Technology for Java® Acceleration.
  - 16 Kbyte Data Cache, 16 Kbyte Instruction Cache, Write Buffer.
  - 220 MIPS at 200 MHz.
  - Memory Management Unit (can therefore run OS such as Linux).
  - Maximum clock speed 240MHz.
  - Embedded Trace Macrocell(ETM)

- This microcontroller is part of the AT91SAM9 series of microcontrollers and has its own dedicated datasheet. It has no on-chip Flash memory, so must boot off external Flash. It has 96k bytes of on-chip SRAM.

- Two External Bus Interfaces (EBIs) (EBI0 and EBI1):
  - EBI0 Supports SDRAM, Static Memory, ECC-enabled NAND Flash and CompactFlash®
  - EBI1 Supports SDRAM, Static Memory and ECC-enabled NAND Flash

- Five 32 bit ports that are multiplexed with on-chip peripheral devices (total of 160 I/O pins):
  - Port A: PA0 to PA31
  - Port B: PB0 to PB31
  - Port C: PC0 to PC31
  - Port D: PD0 to PD31
  - Port E: PE0 to PE31

- LCD Controller
- 2D Graphics Accelerator
- Image Sensor Interface
- 1 × AC97 Controller
- 1 × 10/100 Ethernet MAC
- 2 × USB Full Speed Host Ports
- 1 × USB Full Speed Device Port
- 1 × CAN Controller
- 3 × USARTs
- 1 × DBGU
- 1 × JTAG Port
- 2 × DMA Controllers
- 20 × Peripheral DMA Controller Channels
- 2 × SPI
- 1 × TWI
- 2 × SSC
- 2 × MCI (Multimedia Card Interface)
- 1 × 4 Channel PWM Controller
- 1 × 3 Channel 16-bit Timer/Counter
- 1 × PIT
- 2 × RTT (Real Time Timers)
- 1 × WDT
- 2 × Power-on-reset
- Has no brown-out detection
- 1.2V core voltage
- 3.3V I/O voltage, but can operate the two external bus interfaces at 3.3V or 1.8V, I/O lines of the image sensor can operate at 1.8V, 2.5V, 3V or 3.3V.
- In-system programmable
- Available in TFBGA 324 (324-ball BGA package)

Each I/O pin can sink or source 8mA. A maximum total of 500mA can be drawn from all pins simultaneously.

10.3.2 AT91SAM9263 Evaluation Kit

The AT91SAM9263 ARM9 microcontroller from Atmel can be evaluated using the AT91SAM9263-EK. **Figure 10.3** shows the AT91SAM9263-EK board.

The kit has the following contents:

- AT91SAM9263-EK circuit board
- 12V power supply
- A/B type USB cable
- RS-232 cable with two 9-pin D-type female connectors (crosses tx and rx) lines for connection to a PC.
- RJ-45 crossed Ethernet cable
- 3V CR1225 Lithium Coin Battery
- Stereo in-ear earphones
- CD-ROM

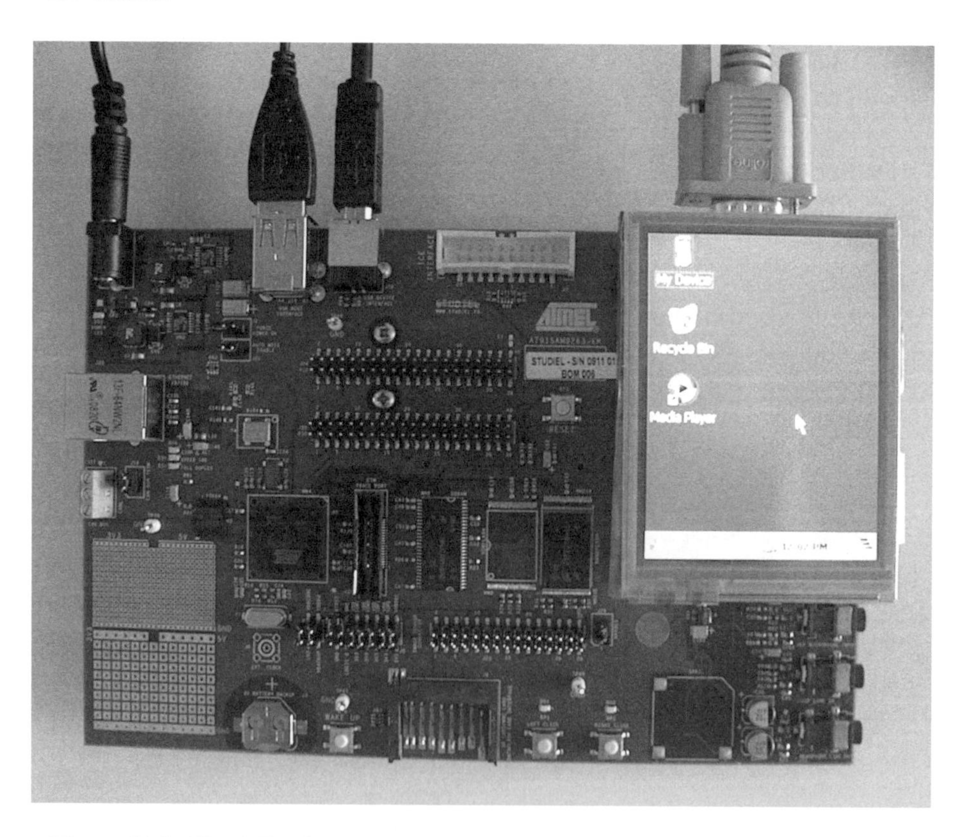

Figure 10.3: The AT91SAM9263-EK Evaluation Board Running Windows CE

The AT91SAM9263-EK board has the following features:

- AT91SAM9263 ARM9 microcontroller
- 64M bytes SDRAM memory on 32-bit data bus (2 × 16-bit chips)
- 4M bytes PSRAM, 16-bit data bus
- 256M bytes of NAND Flash memory with 8-bit data bus
- 1.8” hard disk connector
- TWI serial EEPROM – ATMLH804
- A USB device port
- Two USB host ports
- Two RS-232 serial ports (1 DBGU port, one additional serial port)
- One CAN 2.0B communication port
- JTAG ICE debug interface (20 pin IDC connector)
- One Ethernet 100-base TX with three status LEDs
- AC97 Stereo audio DAC connected to a 3.5mm headphone jack
- One 3.5" 1/4 VGA TFT LCD Module with TouchScreen and backlight
- One ISI connector (camera interface)
- One power LED, two general purpose LEDs
- Two push buttons
- One Wakeup input push button
- A reset push button
- One DataFlash®/SD/SDIO/MMC card slot
- One SD/SDIO/MMC card slot
- Two expansion connectors for unused pins (40 pin headers – 2 rows × 20 pins each)
- A 16.36766MHz crystal for the microcontroller clock

The AT91SAM9263-EK board must be powered with external 12V and does not allow for powering from the USB port. A 12V power supply is supplied with the kit. The centre pin of the power connector is positive.

The Lithium battery must be inserted into its retainer in order for the board to be able to boot up, or the “FORCE POWER ON” jumper J2 must be inserted.

10.3.3 AT91SAM9263 Programming

The AT91SAM9263 can be programmed using the YAGARTO tool-chain.

Demo versions of Windows CE and Linux can be downloaded from **www.at91.com** and run on the board.

The Windows CE demo ran well on the board. It was fast, detected and set up a USB mouse with no problem, detected and set up a USB flash stick and SD card as drives on the system. The media player on the demo played mp3 audio files from the USB flash stick and SD card without a problem. I must say that I was impressed with this demo!

The Linux demo had many more applications available to it and was a little slower to boot.

Both of these operating systems graphic interfaces could be controlled by the touch screen LCD display.

Example programs with source code can be downloaded from the AT91SAM9263 webpage at **www.atmel.com**. The at91sam9263-ek.zip file contains the example programs for the AT91SAM9263-EK board.

Some of the programs from this file were tested and the following should be noted:

1. The make file for each project may have to be modified to change the CHIP and BOARD definitions to at91sam9263.
2. The make file default uses the CodeSourcery toolchain, to use the YAGARTO toolchain, **`arm-none-eabi-`** must be changed to **`arm-elf-`** in the make file. The program compiled and ran successfully using both toolchains.

10.4 ARM Core Comparisons

Table 10.3 compares the three ARM cores of the microcontrollers that have been discussed in this chapter.

Core Type	ARM7	ARM Cortex	ARM9
Processor	ARM7TDMI	ARM Cortex-M3	ARM926EJ-S
Instruction Architecture	32-bit RISC	32-bit RISC	32-bit RISC
Processor Architecture	Von Neumann (ARM v4T)	Harvard (ARMv7-M)	Harvard (ARMv5TEJ)
Instruction Set(s)	ARM 32-bit Thumb 16-bit	Thumb-2 16-bit and 32-bit	ARM 32-bit Thumb 16-bit
Pipeline Architecture	3 stage pipeline - Fetch (F) - Decode (D) - Execute (E)	3 stage pipeline - Fetch (F) - Decode (D) - Execute (E)	- Fetch (F) - Decode (D) - Execute (E) - Data Memory (M) - Register Write (W)
Special Features		- Single cycle 32-bit multiply - Hardware divide	-16 × 32-bit multiplier - DSP Instruction extensions - Java bytecode execution - Instruction and Data Cache - Write buffer - Memory Management Unit (MMU) - Bus Interface Unit

Table 10.3: A Comparison of Three ARM Cores

Index